John Ellor Taylor

Notes on Collecting and Preserving Natural-History Objects

John Ellor Taylor

Notes on Collecting and Preserving Natural-History Objects

ISBN/EAN: 9783337026189

Printed in Europe, USA, Canada, Australia, Japan

Cover: Foto ©berggeist007 / pixelio.de

More available books at **www.hansebooks.com**

NOTES ON

COLLECTING AND PRESERVING

NATURAL-HISTORY OBJECTS.

BY

J. E. Taylor, F.L.S., F.G.S.
F. F. Elwin.
Thos. Southwell, F.Z.S.
Dr. Knaggs.
E. C. Rye, F.Z.S.
J. B. Bridgman.

Professor Ralph Tate, F.G.S.
Jas. Britten, F.L.S.
Professor Buckman, F.G.S.
Dr. Braithwaite, F.L.S.
Worthington G. Smith, F.L.S.
Rev. Jas. Crombie, F.L.S.

W. H. Grattann.

EDITED BY

J. E. TAYLOR, PhD., F.L.S., F.G.S., &c.

NEW EDITION.

LONDON:
W. H ALLEN & CO., 13 WATERLOO PLACE. S.W.

1883.

PREFACE.

The following 'Essays were originally contributed to the pages of 'Science-Gossip,' by the various writers whose names they bear. From the constant queries relating to subjects of this kind, it was deemed advisable to furnish young or intending naturalists with such trustworthy information as would enable them to save time, and gain by the experience of others. For this purpose, the articles have been collected in their present portable form as a Handbook for beginners.

May, 1876.

CONTENTS.

CHAPTER VIII.

CHAPTER IX.

CHAPTER X.

CHAPTER XI.

CHAPTER XII.

CHAPTER XIII.

CHAPTER XIV.

COLLECTING AND PRESERVING.

I.

GEOLOGICAL SPECIMENS.

By J. E. TAYLOR, F.L.S., F.G.S.

THE great end of natural-history reading should be the development of a love for the objects dwelt upon, and a desire to know more about them. This can only be brought about by such practical acquaintance as collecting and preserving them induces. At the same time we should be sorry to see our young readers degenerate into mere collectors! It is a great mistake to suppose, that because you have a full cabinet of butterflies, moths, or beetles, therefore you are a good entomologist; or that you may lay claim to a distinguished position as a geologist, on account of drawers full of fossils and minerals. But this is a mistake into which young naturalists frequently fall. We have seen people with decided tastes for these studies never get beyond the mere collecting. In that case they stand on a par with

B

collectors of postage-stamps. Nor is there much
gained, even if you become acquainted with English,
or even Latin, names of natural-history objects.
Many people can catalogue them glibly, and never
make a slip, and yet they are practically ignorant of
the *real* knowledge which clusters round each object,
and its relation to others. Both Latin and English
names are useful and even necessary; but when you
have simply learnt them, and nothing more, how
much wiser are you than before? No, let the learning of names be the alphabet of science—the means
by which you can acquire a further knowledge of its
mysteries. It would be just as reasonable to set up
for a literary man on the strength of accurately
knowing the alphabet, as to imagine you are a
scientific man the moment you have learned by
heart a few scores of Latin names of plants, fossils,
or insects! Let each object represent so much
knowledge, to which the very mention of its name
will immediately conjure up a crowd of associations,
relationships, and intimate acquaintances, and you
will then see what a store of real knowledge may
be represented in a carefully-arranged cabinet.

The heading of the present articles will have indicated the subject chosen for brief treatment. We
shall never forget the influence left by reading
such charming and suggestive books as Mantell's

'Medals of Creation,' many years ago. Our mind had been prepared for the enthusiasm which this little book produced by the perusal of Page's 'Introductory Text-book,' Phillips's 'Guide to Geology,' and several others of a similar character. But we know of none which impels a young student to go into the field and hammer out fossils for himself, like Dr. Mantell's works. It is impossible not to catch the enthusiasm of his nature. The first place we sallied out to, on our maiden geological trip, was a heap of coal-shale, near a pit's mouth, in the neighbourhood of Manchester. Our only weapon was a common house hammer, for we then knew nothing of the technical forms which geological fancy so often assumes. We had passed that same heap of coal-shale hundreds of times, without suspecting it to be anything more than everybody else considered it viz. a heap of rubbish. Why that particular spot was selected, we cannot now say. We had seen illustrations of carboniferous plants, shells, &c., in books, but we seemed to imagine their discovery could only be effected by scientific men, and that it required a good deal of knowledge before one should attempt to find them. Suffice it to say we made the pilgrimage to the coal-shale heap in pretty much the same mind as we should expect to get the head prize in some fine-art drawing. The humble hammer was

put into use, for a brief time without much effect, as
we could hardly have commenced on a more barren
kind of shale than we had chanced to hit upon.
We imagined we could perceive traces of leaves and
slender stems, but were afraid to trust our eyes. At
any rate, there was nothing definite enough to raise
our enthusiasm. But by-and-by, as the hammer
kept cleaving open the thin leaf-like layers of shale,
there appeared a large portion of that most beautiful
of all fossil plants, the *Lepidodendron*. Those who
are familiar with this object, with its lozenge-shaped
markings running spirally up the stem, will readily
understand the outburst of pleasure which escaped
our lips! That was the first real fossil—a pleasure
quite equivalent to that of landing the first salmon.
How carefully was it wrapped in paper, and carried
home in the pocket! There never was, and never
will be, another fossil in the world as beautiful as
that insignificant fragment of *Lepidodendron*.

We have seen a good many converts made to
geology in a similar manner, since first we laid open
to the light this silent memorial of ages which have
passed away. Let a man have ever so slight acquaintance with geology, and give him the chance of
hammering out a fossil for himself, and the odds are
you thereby make him a geologist for life. There
is something almost romantic in the idea that you

are looking for the first time, and have yourself disentombed the remains of creatures which probably lived scores of millions of years ago! We would strongly advise our readers, therefore, not to fall into the error of supposing that fossil-hunting belongs to highly-trained geologists. On the contrary, it is by fossil-hunting alone that you can ever hope to be a geologist yourself. Another mistake often made, is that of supposing these rich and interesting geological localities are at a distance. It seems so hard to suppose, after reading about typical sections, &c., that under your very feet, in the fields where you have so often played, there occur geological phenomena of no less interest. But it is actually surprising what evidences of our earth's great antiquity, in the shape of fossils, &c., may be studied and obtained in the most out-of-the-way and insignificant places.

You say you have no *rocks* in your neighbourhood —nothing but barren sands, or beds of brick-earth or clay. Well, go to some section of the latter, exposed, perhaps, in some tarn or stagnant pond in a turnip-field. You examine the sides, and what do you see? Nothing, but here and there a boulder-stone sticking out. Well, be content with that. You said you had no rocks in your neighbourhood; how, then, has that boulder, which is a rounded fragment of a rock broken off from somewhere—

how has it come there? Here is a poser at once.
Examine it, and you will perhaps see that its hard
surface is polished or scratched, and then you remember the theory of icebergs, and feel astonished
to think that you hold in your hand an undeniable
proof of the truth of that theory. Those very
scratchings could have been produced in no other
way; that foreign fragment of a rock now only to be
found on some distant mountain-side could have been
conveyed in no other manner. Not content with the
exterior examination, you break the boulder-stone
open, when you may chance to find it is a portion of
silurian, carboniferous or oolitic limestone, and that it
contains *fossils* belonging to one of those formations.
Here is a find—an object with a double interest
turning up where you never expected to discover the
slightest geological incident! You examine other
boulders, and find in them general evidences of
ice-action in their present re-deposition, and most
instructive lessons as to the nature of rocks of various
formations, from the granite and trap series to the
fossiliferous deposits. In fact, there is no place like
one of these old boulder-pits for making oneself
acquainted with petrology, or the nature of stones.

And now, as to the *tools* necessary to the young
geologist. First of all, he cannot take *too few!* It
is a great mistake to imagine that a full set of

scientific instruments makes a scientific man. The following hammers, intended for different purposes, ought to be procured. Fig. 1 is an exceedingly useful weapon, and one we commonly use, to the exclusion of all others. It is handy for breaking off fragments of rock for ex-amination; and, if fossils be included in them, for trimning the specimens for cabinet purposes. As a rule, however, field geologists are always divided over the merits of their hammers, some prefer-ring one shape and some another. Fig. 2 is gene-rally used for breaking up hard rocks, for which the bevel - shaped head is peculiarly adapted. It is usually much heavier

Fig. 1.   Fig. 2.

Pocket Trim-   Duck's-head
ming-hammer.   Hammer.

than the rest, and is seldom used except for specific purposes. If our readers are inclined to study sections of boulder clay, and wish to extract the rounded and angular boulder from its stiff matrix, they cannot do better than use a hammer like Fig. 3. This is sometimes called the " Platypus " pick. Both ends can be used, and the pick end is also good for working on soft rocks, like chalk. A little practice in the field will teach the student how to use these tools, and when, much better than we can describe on paper. The hammers can be obtained from any scientific instrument manufacturer, or from any of the dealers in geological specimens. We have found that the best hammers for usage, however, were to be made out of an old file, softened and well welded, rolled, and then hammered into a solid mass. If properly tempered a hammer made in this fashion will last you your life.

Fig. 3.

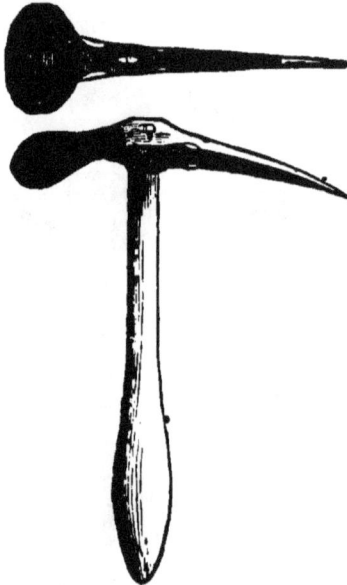

" Platypus " Pick for clay, &c.

So much for the rougher weapons of geological strife. Next, be sure and provide yourself with *thick-soled* shoes or boots. Geological study will take you into a good many queer places, wet and dry, rough and smooth, and it is absolutely necessary to be prepared for the worst. Patent leather boots and kid gloves are rarely worn by practical geologists. And we have heard it remarked at the British Association meetings, that they could always tell which members belonged to the Geological Section by their *thick-soled* boots. A similar remark applies to clóthes. The student need not dress for the quarry as he would for the dining room. Good, strong, serviceable material ought to be their basis.

Secondly, as to the student's comforts and necessaries. These are generally the last thing an ardent naturalist thinks about. For ourselves, however, we give him ample leave to provide himself with pipe and tobacco, should his tastes lie in that direction. *We* never enjoyed a pipe half so much as when solitarily disinterring organic remains which had slumbered in the heart of the rock for myriads of ages. As to the *beer*, we can vouch that it never tastes anything like so good as during a geological excursion.

We have found the leathern bags sold for school-book purposes to be as handy to deposit specimens in, during a journey, as anything else. They have the merit of being cheap, are strong, and easily carried. If not large enough, then get a strong, coarse linen havresack, like that worn by volunteers on a field day. Paper, cotton *wadding* (not wool), sawdust for fragments of larger fossils, intended to be repaired at home, wooden pill-boxes, and a few boxes, which may be obtained from any practical naturalist, with *glass tops*, are sufficient "stock-in-trade" for the young geologist. The wadding does not adhere to the specimens as wool does, and the glass-topped boxes are useful, as it is not then necessary to open a box and disinter a delicate fossil from its matrix in order to look at it. Add a good strong pocket lens, such as may be bought for half-a-crown, and your equipment will be complete. If you intend to study any particular district, get the sheets published by the Geological Survey. These will give you, on a large scale, the minute geology of the neighbourhood, the succession of rocks, faults, outcrops, &c. In fact, you may save yourself a world of trouble by thus preparing yourself a week or so before you make your geological excursion. The pith of these remarks applies with equal force if you purpose, first of all, to examine the neighbourhood in which you live. Don't do so until you have read all that has been written about it, and examined all the available maps and sections. This advice however, applies more particularly to *geological* examination of strata. If you are bent chiefly on *palæontological* investigation, that is, on the study of *fossils*, perhaps it will be best just to read any published remarks you may have access to, and then boldly take the field for yourself. In addition to a hammer, we would advise the young student to take a good narrow-pointed steel chisel, and a putty-knife. The former is very useful for working round, and eventually obtaining, any fossil that may have been weathered into relief. The latter is equally serviceable for clayey rocks or shales.

In arranging the spoils of these excursions for the cabinet, a little care and taste are required. We will suppose you to possess one of those many-drawered cabinets which can now be obtained so cheaply. Begin at the bottom, so that the lowest drawers represent the lowest-seated and oldest rocks, and the uppermost the most recent. If possible, have an *additional* cabinet for *local* geology, and never forget that the first duty of a collector is to have his own district well represented! A compass of a few miles will, in most cases, enable him to get a store of fossils or minerals which cannot well be obtained elsewhere. Supposing he is desirous of having the geological systems well represented, he can always do so by the insertion of such paragraphs as those which appear in the Exchange columns of 'Science Gossip.' It is by well and thoroughly

working separate localities in this fashion that the
science of geology is best advanced. You hear a
good deal about the "missing links," and it is an
accepted fact that we, perhaps, do not know a tithe
of the organic remains that formerly enjoyed life
Who knows, therefore, but that if you exhaust your
district by the assiduous collection of fossils, you
may not come across such new forms as may settle
many moot points in ancient and modern natural
history? The genuine love of geological study is
always pretty fairly manifested in a student's
cabinet. Science, like charity, begins at home.
It impels a man to seek and explain that which
is nearest to him, before he attempts the elucidation
of what really lies in another man's territory!

It is not necessary that the student should waste
time in the field about naming or trying to remember the names of fossils, &c., on the spot. That can
be best done at home, and the pleasure of "collecting" can thus be spun to its longest length. Box
them, pack them well (or all your labour is lost),
and name them at home. Or supposing you do
not possess books which can assist you in nomencla-
ture, carry your fossils or minerals, just as you
found them, to the nearest and best local museum,
where you will be sure to see the majority of them
in their proper places and with their proper names.

Copy these, and when you arrange your specimens in the cabinet, either get printed cards with the following headings—

*Genus*_____

Species _____   _____

Formation _____ _

*Locality*_____

(which can always be obtained at a cheap rate from the London dealers), or else set to work and copy them yourself in a good plain hand, so that there is no mistaking what you write. As far as possible, in each drawer or drawers representing a geological formation, arrange your specimens in natural-history order—the lowest organisms first, gradually ascending to the higher. By doing so, you present geological and zoological relationship, so that they can be taken in at a glance. You further make yourself acquainted with the relations of the fossils in a way you never would have done, had you been content to huddle them together in any fashion, so that you had them all together. Glass-topped boxes, again, are very useful in the cabinet, especially for delicate or fragile fossils, as people are so ready to take them in their hands when they are shown, little thinking how soon a cherished rarity may be destroyed, never to be replaced. Pasteboard trays, made of stiff green paper, squared by the student according to size, can also be so arranged as that the drawer may be entirely filled, and so the danger of shaking the contents about may be removed. Each tray of fossils ought to have the above-mentioned label fastened down in such a way as that it cannot by accident get changed by removal.

The spring and summer time are fast approaching, and we know of nothing that will so much assist in their rational enjoyment as the adoption of some study in natural science. Botany, entomology, ornithology, geology, are all health-affording, nature-loving pursuits. We have passed some of the very happiest moments of our lives in solitary quarries, or on green hill-sides,

"The world forgetting, by the world forgot!"

There, amid the wreck of former creations, and with the glory of the present one around us, we have yielded to the delicious sense of reverie, such as can only be begotten under such circumstances. The shady side of the quarry has screened us from solar heat, and, whilst the air has been melodious with a thousand voices, we have made personal acquaintance with the numerous objects of God's

·creation, animals and plants. How apt are the thoughts of the poet Crabbe, and how well do they convey the feeling of the young geologist in such places:

> "It is a lonely place, and at the side
> Rises a mountain rock in rugged pride;
> And in that rock are shapes of shells, and forms
> Of creatures in old worlds, and nameless worms;
> Whole generations lived and died, ere man,
> A worm of other class, to crawl began."

II.

BONES.

By Edward Fentone Elwin, Caius College.

Why is it that the students of Osteology are so few in number? It is a branch of science which offers a wide field for original research, and one in which at every step one's interest must get more and more engrossed. It is a branch of science in which a sufficient portion of its elements may be rapidly learned, in order to set the student fairly on his road. The barriers which surround it are few: that is to say, the *technical* barriers are few. Many people who want to occupy themselves with scientific study are deterred, because of the feeling that there are so many laborious preliminaries to be gone through before they can begin to take any real pleasure in the pursuit. Now, in Osteology it is true that a wide and really almost unexplored field lies open before one, but the equipments necessary to fit one for one's journey are easily attained. The first step is to get thoroughly acquainted with some one typical specimen, as a standard of comparison for all future work. It matters little what species is taken;

whichever comes most convenient. Some familiar mammal of fair size is the best. The dog is as good as any, and easy to obtain. There ought never to be any real difficulty in getting a suitable specimen. If expense is nó object, the simplest way is to get a preparation, set up so as readily to take to pieces, at any of the bone-preservers' shops in London. One like this costs only a moderate sum, and is, of course, the least trouble, although the manner in which professionals prepare their bones is not altogether satisfactory. But we may regard this as something in the light of a luxury ; and it is not hard to prepare one's own specimens, provided we do not mind a little manipulation with unsavoury objects. I have given hints as to the best method by which this may be done in various pages of 'Science-Gossip.' * Of course, as one's work gets on, one needs further specimens, but I do not think that anyone who keeps his eyes open need be at a loss in this matter. I have picked up several admirable bones ready cleaned by the wind and weather, and many slightly damaged ones may be got at naturalists' shops for small sums, which are almost as good as the perfect ones for an observer's purposes. Even single and isolated bones are often very instructive.

But the first main point is that of getting the

* 'Science-Gossip' for 1873, p. 39 ; for 1874, p. 226.

C

forms, peculiarities, names, and positions of the
bones of one skeleton fully impressed on the student's
mind. As to the books which are to help him to do
this, it is very hard to know what to recommend.
As far as I know, there is no really luminous book
on osteology in existence. So far as learning the
names and peculiarities of the bones, nothing could
be better or more to the purpose than Flower's
' Osteology of the Mammalia '; but this treats only
of one class, and does not get beyond technical
description. The first and second volumes of
Owen's ' Comparative Anatomy of Vertebrates '
fill the gap the best of any, and yet these are by no
means what we really want. There is a good deal
about bones in Huxley's ' Anatomy of Vertebrated
Animals,' but in such a fragmentary and scattered
form as to be of little use. The fact is, the field is
yet open for an Osteological Manual. Much has
been written on the subject. Pages of precise and
accurate description, beautiful and artistic sheets of
plates of bones without number, can be seen in any
scientific library. But this is only half the matter.
We want to advance a step farther. It is the
relation between structure and function which needs
working out.

When a new bone finds its way into the student's
hands, he observes some peculiarity in shape or

structure in which it differs from the bones he is already acquainted with; the question naturally occurs to him, Why does this bone assume one shape in one animal, and in another is modified into a different form? He may look in vain in his books for an answer to his query. And yet it is points like these which, in my opinion, make up the true science of Osteology. It is through careful, constant, and intelligent observation, that these enigmas are to be solved. Observation, indoors and out; close attention to the habits of the animal in question, on the one hand, and careful consideration of its anatomical peculiarities, on the other.

Let me give an instance of this, first of all taking it as an axiom that everything has been done with a purpose. Take, then, the skull of a crocodile. What do we find? The orbits of the eyes, the nasal orifice, the passages leading to the auditory apparatus, all situated on a plane, along the upper flattened surface of the head. What, then, is the cause of this? Palpably to allow the crocodile to remain submerged in the water, with its nose, eyes, and ears just above the surface to warn him of the approach of enemies or prey, and the rest of his carcase securely hidden beneath the waters.

Take another instance. Observe the habits of a mole. With what rapidity it burrows underground, shovelling away the earth with its fore feet. Then look at its skeleton. We find just what we should have expected. The bones of its fore legs of astounding strength and breadth, furnished with deep grooves, which, together with its sternum or breastbone, which is furnished with a keel almost like that of the sternum of a bird, afford attachment to the powerful muscles. Its hind legs, being sim- ply needed for locomotion, are of the normal size. So, also, with the birds. The size of the keel of the sternum varies in proportion to the powers of flight which each species requires, for it is to the broad surfaces of the sternum that the great wing-muscles are attached. Take the skeleton of a humming- bird, which spends its life almost upon the wing. We find there a keel of so vast a size, that the re- mainder of the skeleton is reduced to insignificance in comparison. Of course, these instances that I have given are all of the most obvious nature, but they serve to show my meaning; and the same line of reasoning can, I am sure, be extended to all the more minute points in osteological structure.

In these researches, one is soon struck by the fact that in the modifications in various bones, or sets of bones, in accordance with the habits of each animal, the original type is never departed from, only modified. See, for example, the paddle of a whale. More like the fin of a fish in general appearance, and yet the same set of bones which are found in the arm of a man, are again found in an adapted form in the paddle of the whale. So, also, the fore leg of a horse preserves the same general plan. What is generally called its knee is in reality its wrist. It is there that we find the little group of bones which forms the carpus. All below it answers to our hand—a hand consisting of one finger.

Take even a wider instance. Compare the arm of a man and the wing of a bird. Still greater adaptations have taken place, and yet the plan remains the same. We still find the clavicle or collar-bone, the scapula or shoulder-blade, the humerus, ulna, and radius, answering to the same bones of our arm, a small carpus or wrist, and finally the phalanges or fingers, simplified and lengthened and anchylosed to form but one series of bone, with the exception of a rudimentary thumb. It is not uncommon to find a rudimentary bone like this which in some allied species is fully developed. The leg of the horse again gives us a very striking example of this. There is, so to speak, only a single finger, but we find, one on each side of this single finger, two small bones, commonly known only as splint-bones. These are the rudimentary

traces of the same finger-bones, which in the rhino-
ceros are fully developed.

Now Osteology abounds in wonderful forms of
structure like these. It is a study pregnant with
pleasurable results, and is a real profitable study,
and one in which each fresh student may do real
solid work. It is all the little facts observed by
naturalists from time to time all over the world,
which on being collected together form the nucleus
of knowledge; for indeed all the scientific knowledge
which we possess is little more than a nucleus, with
which we are supplied. The mere collector of
curious objects in no way furthers science. Plenty
of people have amassed beautiful collections of
insects interesting in their way, but of very tran-
sient interest if it goes no farther. The collector
possibly knows nothing at all of the wonderful
internal structure of the animals he preserves. His
insects are to him simply a mosaic—a collection of
pretty works of art. So also the shell-collector—
for I cannot call such a one as I describe a concho-
logist—has often, I believe, the most vague ideas of
what kind of animals they were that dwelt in the
cases he so carefully treasures, and his collection is
consequently of a dubious worth to him. Now, to
those who study the anatomy of the mollusc as
well as its shell, such a collection is full of the

deepest interest. He has learnt from his dissections that the habits of every variety of mollusc are accompanied by a variety of structure, which occasions a variety in the shape of the case which envelopes it. It all blends together, and forms a harmonious whole. With a real love for science, as doubtless some of these collectors have, one is sorry to see so much time and money wasted on a pursuit which in their hands yields no fruit of any worth. The work of the mere collector can only be classed with that of the compiler of a stamp-album. Whereas, collections of natural objects, combined with intelligent study, are invaluable and almost indispensable to the naturalist.

In Mr. Chivers's note on Preserving Animals, No. 117 of 'Science-Gossip,' the following passage occurs:—"The skeleton must be put in an airy place to dry, but not in the sun or near the fire, as that will turn the bones a bad colour." I cannot comprehend how this idea should have arisen. Perhaps the most indispensable assistant to the skeleton preparer is that very sun which Mr. Chivers warns him against. The bleaching power of the rays of a hot summer sun is astounding, and bones of the most inferior colour can rapidly be turned to a beautiful white by this means. It is for want of time and care in following out this method that the professional

skeleton preparers in London resort to the aid of
lime, which, although it makes them white, is ter-
ribly detrimental to the bones themselves. In a
smoky city like London, the principle of sun-
bleaching would be hard to follow; but so great is
its value, that more than once I have had valuable
specimens sent down to me in the country, by a comparative anatomist in London, to undergo a course of
sun-bleaching; and a specimen which I have received stained and blotched, I have returned of a
beautiful uniform white, a change entirely due to
that sun which we are told to beware of.

The question, How are skeletons to be prepared?
is one which is repeatedly asked. People desire a
method by which with little trouble the flesh may
be removed from a specimen, and a beautiful skeleton
of ivory whiteness left standing in its natural position. I can assure all such inquirers that this cannot be accomplished by any method at all. The art
of preparing bones is a long, elaborate, and difficult
one, and he who wishes to become a proficient in it
must be alike regardless to the most unpleasant
odours, and to handling the most repulsive objects.
Mr. Chivers's receipt for the maceration of specimens
is about the best which one could have, only I should
not advise so frequent a change of the water. What
is needed is as rapid a decomposition of the flesh as

is possible, and then the cleaning of the skeleton just before the harder ligaments have also dissolved But this requires very careful watching, and with the utmost pains it is almost impossible to get a skeleton entirely connected by its own ligaments.

Another point which must be taken into consideration is this: What use is to be made of the specimens after they are prepared? Are they for purposes of real study, or simply as curious objects to look at? If the latter is the purpose, I must confess I do not think they are worth the trouble of preparing. If the former is the object for which they are intended, then I think no care or pains are thrown away. But for the real student of Osteology the separated bones, as a rule, are far more valuable than those which are connected. He needs one or two set up for purposes of reference, but the great bulk of his specimens should be separate bones. Osteology is one of the most delightful branches of comparative anatomy, and one not very hard to master. Let anyone try the experiment by getting together a few bones—and those from the rabbit or the partridge we have had for dinner are by no means to be despised—and then, by purchasing Flower's 'Osteology of the Mammalia,' which is a cheap and first-rate book, he will learn what the study of the skeleton really is. And then let him

be on the look-out for specimens of all kinds on
all occasions, bringing home all suitable objects he
meets with in his walks, however unsavoury they
may be, and he will be astonished to find how many
specimens he will get together in the course of a
year. I have now myself upwards of seventy skulls
of various kinds, with often the rest of the skeleton
as well, the greater part of which were gradually
collected, by keeping constantly on the watch for
them, within a year and a half.

III.

BIRDS' EGGS.

By Thomas Southwell, F.Z.S., etc.

I can imagine no branch of natural history more fascinating in its nature, or more calculated to attract the attention of the young, than the study of the nests and eggs of birds; the beauty of the structure of the one, and of the form and colour of the others, cannot fail to excite wonder and admiration; and the interest thus excited, if rightly directed, may, and indeed has, in many instances, lead to the development of that passionate love for all nature's works, that careful and patient spirit of investigation, and that deep love for truth which should all be characteristics of the true naturalist. Who can look back upon the days, perhaps long passed away, when as a school-boy he wandered through the woods and fields, almost every step unfolding to him some new wonder, some fresh beauty—glimpses of a world of wonders only waiting to be explored—who can look back to such a time without feeling that in those wanderings there dawned upon his mind a source of happiness which in its purity and intensity ranks

high amongst those earthly pleasures we are per-
mitted to enjoy, and which has influenced him for
good in all the changes which have since come upon
him, lightening the captivity of the sick room, and
adding fresh brightness to the enjoyments of health.

Between the true naturalist and the mere "collector" there is a wide gap, and I trust that none for
whom I am writing will allow themselves to drift into
the latter class; the incalculable mischief wrought by
those who assist in the extermination of rare and
local species by buying up every egg of a certain
species which can be obtained, for the mere purpose
of exchange, cannot be too much deprecated, and I
hope that none of my readers will be so guilty; to
them the pleasures of watching the nesting habits
of the bird, the diligent search and the successful
find are unknown; the eggs in such a cabinet are
mere egg-shells, and not objects pregnant with interest, recalling many a happy ramble, and many
a hardly-earned reward in the discovery of facts and
habits before unknown. Every naturalist must be
more or less a collector, but the naturalist should
always be careful of drifting into the collector, his
note-book and his telescope should be his constant
and harmless companions.

When the writer first commenced his collection,
the mode of preparing the specimens for the cabinet
was very rude indeed, and the method of arranging

equally bad; he is sorry to say the popular books upon the subject which he has seen do not present any very great improvement; in giving the results of his own experience, and the plan pursued by the most distinguished oologists of the day, who have kindly allowed him to explain the methods they adopt, he will, he trusts, save not only much useless labour, but many valuable specimens.

Before saying a word as to preparing specimens for the cabinet, I wish to impress upon the young oologist the absolute necessity for using the greatest care and diligence in order satisfactorily to identify, beyond possibility of doubt, every specimen, before he admits it to his collection. Without such precautions, what might otherwise be a valuable collection is absolutely worthless; and it is better to have a small collection of authentic specimens than a much larger one, the history of which is not perfectly satisfactory; in fact, it is a good rule to banish from the cabinet every egg which is open to the slightest doubt. There are some eggs which, when mixed, the most experienced oologist will find it impossible to separate with certainty, and which cannot be identified when once they are removed from the nest.

The difficulties in the way of authentication are by no means slight, but space will not allow me to dwell upon them; the most ready means, however, is that of watching the old bird to the nest, although even in this, as the collector will find by experience, there is a certain liability to error. In collecting abroad it will be found absolutely necessary (however reluctant we may be to sacrifice life) to procure one of the parents with the nest and eggs. As we are writing for beginners at home, we trust such a measure will rarely be necessary; but that an accurate knowledge of the appearance of the bird, its nesting habits, the situation, and the materials of which the nest is composed, will be found amply sufficient to identify the eggs of our familiar birds. This knowledge of course is only to be obtained by patient and long observation; but it is just by such means that the student obtains the practical insight into the habits and peculiarities of the objects of his study, together with the careful and exact method of recording his observations, which eventually enables him to take his place amongst the more severely scientific naturalists whom he desires to emulate.

I will first describe the tools required, and then proceed to the mode of using them.

Figs. 4 and 5 are drills used for making the hole in the side of the egg, from which the contents are discharged by means of the blowpipe, Fig. 6. Fig. 4 has a steel point, brass ferrule, and ebony handle, and may be used for eggs up to the size of the wood-pigeon's; Fig. 5 is all steel, the handle octagonal, to give a firm hold to the fingers in turning it, and may be used for eggs from the size of the wood-pigeon's upwards. The points of both are finely cut like the teeth of a file, as shown in the woodcut. The blowpipe, Fig. 6, is about 5½ inches in length (measured along the curve), and is made of German silver, which from its cleanliness, lightness, and freedom from corrosion, will be found the most suitable: it should be light and tapering, and with a ring at the upper eid to prevent it from slipping out of the mouth when used. A piece of thin wire, Fig. 7, should be kept in the tube when not in use, to prevent it from becoming stopped up by any foreign substance. A common jeweller's blowpipe may be used for large eggs, such as those of gulls and ducks. Fig. 8 is a small glass bulb-tube, which may be used for sucking out the contents of very delicate eggs, and other purposes, which will be explained here-

Fig. 4.   Fig. 5.

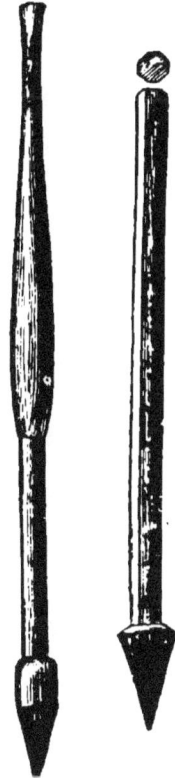

Drills for perforating Birds' Eggs.

Fig. 6.　　Fig. 7.

Fig. 8.

German-silver
Blowpipe.

Wire for unstop-
ping ditto.

Glass Bulb-tube,
for sucking eggs.

after. The small drill and blowpipe may be carried inside the cover of the note-books.

The sooner a fresh egg is emptied of its contents after it is taken from the nest the better. This should be done by making a hole in the side with the drill (choosing the side which is least conspicuously marked) by working it gently backwards and forwards between the forefinger and thumb, and taking great care not to press too heavily, or the egg will burst with the outward pressure of the drill : a very small hole will generally be found sufficient. When this is done, take the egg in the left hand with the hole *downwards*, introduce the blowpipe, by blowing gently through which, the contents may soon be forced out. Water should then be introduced by means of a syringe or the bulb-tube, which may be filled and blown into the egg. After shaking, blow the water out again by means of the blowpipe ; repeat this till the egg is free from any remains of the yolk or white : should the egg not be quite fresh, it will require more washing. Care should be taken to wet the surface of the egg as little as possible. After washing the interior, lay the egg, with the hole downwards, on a pad of blotting-paper to drain till it is quite dry. Should the eggs be much incubated, I should recommend that the old birds be left to complete their labour

D

of love; but a valuable egg may be made available by carefully cutting a piece out of the side, extracting the young one, and, after replacing the piece of shell with strong gum-water, covering the join with a slip of very thin silk-paper, which may be tinted so as to resemble the egg, and will scarcely be noticed. This is a very rough way of proceeding, however, compared with Professor Newton's plan of gumming several thicknesses of fine paper over the side of the egg to strengthen it, through which the hole is drilled: the young chick is then cut into small pieces by means of suitable instruments, and the pieces removed with others:* the paper is then damped and removed from the egg.

The old plan of making two holes in the side of the egg is very objectionable: a hole at each end is still worse. Many eggs would be completely spoiled by washing; none improved. There is no necessity for washing at all, except such as are very filthy, and these eggs (which you may be sure are not fresh) are not such as should be willingly accepted as specimens: a *little* dirt only adds to the natural appearance of the egg; washing in most cases cer-

---

* "Suggestions for forming Collections of Birds' Eggs." By Professor Newton. Written for the Smithsonian Institution of Washington, and republished by Newman, 9, Devonshire Street, Bishopsgate.

tainly does not. Never use varnish to the shell; it
imparts a gloss which is not natural : all eggs should
not have a polished appearance like those of the
Woodpecker. Should the yolk be dried to the side
of the egg, a solution of carbonate of soda should
be introduced : let it remain till the contents are
softened, then blow out and wash well. Great care
must be taken not to allow the solution to come in
contact with the outside of the egg. Having blown
the egg, and allowed the inside to become quite dry,
procure some thin silk-paper gummed on one side,
and with a harness-maker's punch cut out a number
of little tickets suitable to the size of the hole in the
egg, moisten one of these, and place it with the gum
side downwards over the hole, so as to quite cover
it; cover the ticket with a coat of varnish, which
will render it air-tight and prevent its being affected
by moisture. The egg thus treated will have all the
appearance of a perfect specimen, and if kept from
the light will suffer very little from fading.

The note-book has been mentioned. This should
be a constant companion ; nothing should be left to
memory. When an egg is taken, a temporary pencil
number should at once be placed upon it, and this
number should correspond with the number attached
to an entry in the note-book, describing the nest (if
not removed), its situation, number of eggs, day of

month, and any other particular of interest. When the egg is ready for the cabinet, as much of this information (certainly, name, date, and locality) should be indelibly marked upon it as conveniently can be done (neatly, of course, and on the under side); also the number referring to the collector's general list of his collection, into which the important parts of the entry from the note-book should be copied. Never trust to gummed labels, which are always liable to come off; by writing the necessary particulars upon the egg itself there can be no confusion or mistake. Most collectors have their own plan of cataloguing their collection. I have adopted the following, which I find to answer very well. Obtain a blank paper book the size of common letter-paper, rule a horizontal line across the centre of each page, and make a complete list of British birds, placing only two names on each page, one at the head of each division, prefixing a progressive number to each name : this number is to agree with that marked on the egg of the species named. Then follow the locality whence the egg came, by whom taken (if not by myself), or how it came into my possession, with any other particular worthy of note. With all eggs received in exchange or otherwise, this note should, if possible, be obtained in the handwriting of the person from whom they are re-

-ceived, and the slip on which it is written be affixed in the book under the number. When specimens of the eggs of the same species are obtained from various localities, those from each locality should be distinguished by a letter prefixed to the number. The plan will be better understood by referring to the following extract :

62. GREAT SEDGE-WARBLER (*Sylvia turtoides*, Meyer).

62. Received of ——, from the cabinet of Mr. ——.

*a*62. Taken by ——, a servant of ——, on the banks of the river Tongreep, near Valkenswaard, in the south of Holland, on the 9th of June, 1855. The birds may be heard a long way off by their incessant "Kara, Kara, Kara." A few years ago not one was to be found near Valkenswaard.

A. B——

*b*62. Bought at Antwerp in August, 1865.

118. MEALY RED-POLE (*Fringilla borealis*, Tem.).

118. Nyborg, at the head of Mæsk Fjord (one of the two branches into which Varanger Fjord divides), East Finmark, Norway, July, 1855. The birds were very plentiful, and only one species seen, which appears quite identical with that which visits England every winter.

C. D. E——

By means of these entries, and the corresponding number on the egg, mistakes are impossible, and the name and history of each egg would be quite as well known to a stranger as to the possessor. It needs not to be said that this catalogue is replete with the deepest interest to its compiler. In it he sees the record of many a holiday trip and many a successful find. Some of the entries in my own register —the earliest date back five-and-twenty years—are memorials of companions long since dead, or separated by rolling oceans, but on whose early friendship it is a pleasure to dwell.

Nothing can be more vexatious and disappointing than the receipt of a box of valuable eggs in a smashed or injured condition from want of care or knowledge of the proper method of packing. A simple method is recommended by Professor Newton, which, from experience, I can confidently recommend:—Roll each egg in tow, wool, or some elastic material, and pack them closely in a stout box, leaving no vacant place for them to shake; or a layer of soft material may be placed at the bottom of the box, and upon it a layer of eggs, each one wrapped loosely in old newspaper; upon this another layer of wool or moss, then again eggs, and packing alternately until the box is quite full. Bran, sawdust, &c., should never be used; and it should be ascertained that the box is quite filled, so that no shaking or settlement can occur.

Almost every collector has his own plan for constructing his cabinet, and displaying his collection

The beginner, if left to himself, will find it a matter of no small difficulty, and many will be the changes before he arrives at one at all satisfactory. Mr. Osbert Salvin has invented a plan which I think as near perfection as it is possible to arrive at, and through his kindness I am enabled to give a brief description of it. In the first place, his cabinets are so constructed that the drawers, of different depths, are interchangeable. This is effected by placing the runners, which carry the drawers, at a fixed distance from each other and making the depth of each drawer a multiple of the distance between the runners. For example: if the runners are three-quarters of an inch off each other, then let the drawers be $1\frac{1}{2}$, $2\frac{1}{4}$, 3, $3\frac{3}{4}$, $4\frac{1}{2}$, &c., inches deep. All these drawers will be perfectly interchangeable, and a drawer deep enough to hold an ostrich's egg can in a few moments be placed amongst those containing warblers': every requirement of expansion and rearrangement will be vastly facilitated, involving none of those radical changes so worrying to a collector.* Mr. Salvin's plan of arranging the eggs is equally simple, and admits of any amount of change with very little trouble. Each drawer is divided longitudinally by

---

* Of course, cabinets thus constructed will be found equally convenient for collections of bird-skins, fossils, &c.

thin slips of wood into three or more parts, about
4 to 6 inches across, as may be convenient; a number
of sliding stages are then constructed of cardboard,
by cutting the cardboard half through, at exactly
the width of the partition, and bending the sides down
to raise the stage to the required height. A section
of one of these stages will be seen in Fig. 9, and the

Fig. 9.

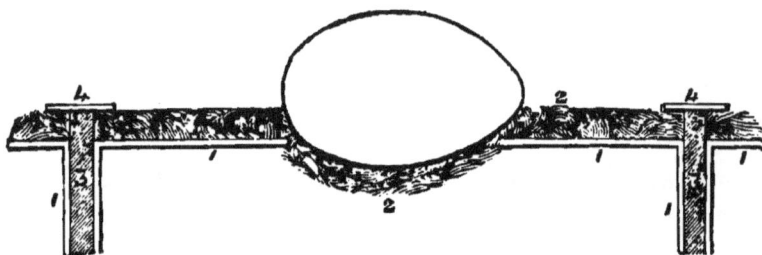

Section of Sliding Stage.

arrangement in the drawer at Fig. 10. A number of
oval holes are then to be cut by hand, or with a
wadding-punch of suitable size (altered in shape by
hammering), and a thin layer of cotton-wool gummed
on the upper surface of the stage: the holes, of
course, should be suitable in size to the egg they are
intended to receive. Between these stages sliding
partitions must be placed : these should be made of
just sufficient height that the horizontal part may
fit closely on the wool, as shown at Fig. 9. These
partitions should be made of thin wood for the

upright part, along which a horizontal strip of cardboard is to be fastened with glue, on which is to be
placed a label bearing the name of the species of

Fig. 10.

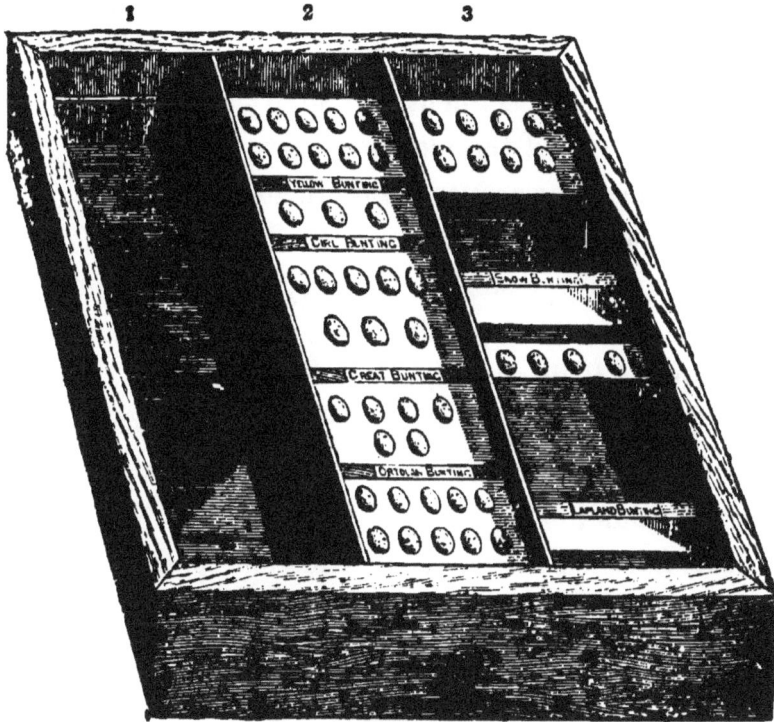

Cabinet Drawer on Mr. Salvin's plan.

egg displayed on the stage, as seen in Fig. 10. All
this will be better understood by referring to the
figures.

Fig. 9 represents a longitudinal section of one of

the stages in its place, with the ends of the two next; 1, showing the cardboard stage; 2, the cotton-wool; 3, the sliding partition; and 4, the horizontal slip of cardboard to carry the label.

Fig. 10 represents one of the drawers on Mr. Salvin's plan: it is divided into three parts (1, 2, 3) by fixed partitions. No. 1 is represented empty; No. 2 with the specimens arranged; No. 3 with two stages and two of the movable partitions.

This may appear very complicated at first sight, but a few trials will be sufficient to master the details, and the result will be very beautiful if neatly carried out. The eggs are well shown, not liable to fall out of their places, and it is very little trouble to alter the arrangement, every part being movable. Each drawer should be covered by a sheet of glass to exclude dust.

Mr. Salvin's cabinet is an excellent one for holding the nests of birds, which should be removed with as little damage as possible, and placed in the drawers, under cover of glass. Great care must be taken to keep them free from moth, to which they are very liable; for this purpose they should be dressed with the solution of corrosive sublimate.

The young collector should remember that what is worth doing at all is worth doing well, and that the care bestowed upon his cabinet is not labour in

vain ; habits of exactness and precision of arrangement are absolutely necessary if he would make the
best use of the materials which come in his way;
and, above all, never let him degenerate into the
mere collector : his collection should be for use, and
not merely ornamental.

IV.

BUTTERFLIES AND MOTHS.

By Dr. HENRY KNAGGS.

THE collector of Lepidoptera who aspires to success must read the book of nature as he runs. If he have not the wit to note and turn to account each little fact which may come under his observation, neither he nor science will be the better for his collecting. He should, whenever he makes a capture, *know the reason why*, or he will never make a successful hunter. He should be ever on the alert: his motto, *nunquam dormio*.

Some collect for profit, others for pastime; but the aim of our readers, I take it, is not only to acquire a collection of really good specimens, but also at the same time to improve their minds; and the best way of effecting this purpose is to hunt the perfect insect, not so much for itself as for the sake of the golden eggs, which, with proper care and attention, will in due course yield the most satisfactory results in the shape of bred specimens.

This being the case, and space being limited, it

seems best to simply touch upon the preliminary
stages of insect existence, pointing out as we go
those methods of collecting and preserving which
experience has shown to be the most successful.

There can be no doubt but that egg-hunting is a
very profitable occupation, and far more remunerative than most people dream of, particularly as a
means of acquiring the Sphinges, Bombyces, and
Pseudo-bombyces. Eggs, speaking generally, are
to be found on the plants to which the various
species are attached; and a knowledge of the time
during which the species remains in the egg state,
as well as the appearance of the eggs as deposited
in nature, should if possible be acquired previous to
proceeding to hunt. The most practical way of ascertaining the food and time is to watch the parent
insect in the act of depositing her ova; but when
the plant has been thus discovered, the best way is
to capture her, and induce her to lay at our home.
When eggs are inconspicuous, of small dimensions,
or artfully concealed, the use of a magnifying glass
is invaluable.

Eggs may be preserved by plunging them in
boiling water or piercing them with a very fine
needle, or they may have their contents squeezed
out and be refilled by means of a fine blowpipe, with

some coagulable tinted fluid ; but the shells themselves, after the escape of the larvæ, form, when
mounted, beautiful objects for the microscope.

The three most successful plans of obtaining
caterpillars are searching, beating, and sweeping.
The first requires good eyesight and a certain
amount of preparatory knowledge ; the others are a
sort of happy-go-lucky way of collecting, useful
enough and profitable in their way, but affording a
very limited scope for the exercise of the wits. In
searching for larvæ, the chief thing is to observe the
indications of their presence. A mutilated leaf, a
roughened bark, a tumid twig, a sickly plant, an
unexpanded bud, an abortive flower, or a windfall
fruit, should at once set us thinking as to the cause ;
or, again, the webs, the silken threads, the burrow-
ings and trails, or the cast-off skins of larvæ, may
first call our attention to their proximity. Of course,
larvæ may be found on almost all plants, as well as
in the bark, stems, or wood of many ; but the collector should fortify himself with a knowledge of
what each plant is likely to produce, and hunt ac-
cordingly ; for though indiscriminate collecting may
sometimes be successful, it does not tend to improve
the intellectual powers.

Beating is the more applicable method of working
trees and bushes. It is carried out by jarring the

larvæ from their positions by the aid of a stick or pole, in such a manner that they will fall into an inverted umbrella, or net ; or a sheet may be spread beneath for their reception. Sweeping with a strong net, passed from side to side with a mower-like movement, is better adapted for working low ground-herbage. The umbrella net, shown in Fig. 11, is, perhaps, the best for the purpose. It is constructed

Fig. 11.

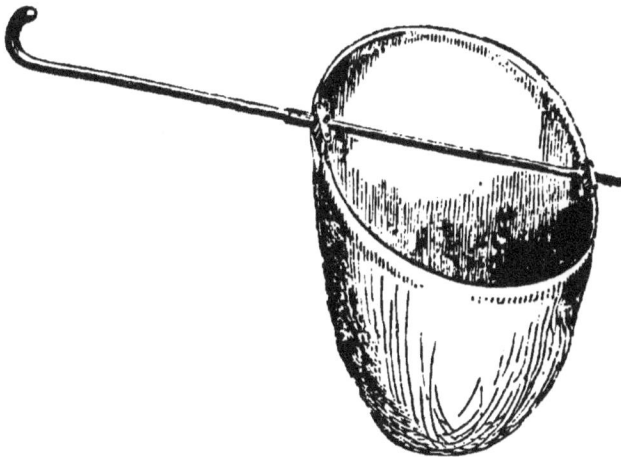

Umbrella Net.

by hinging two lengths of jack-spring on two pieces of brass, and adapting them to the stick of the net, the upper piece of brass being fixed, the lower movable.

When captured, larvæ should be transferred to

chip boxes, or else to finely and freely perforated tins, the latter better preserving the food. A very handy box for the purpose is formed by fitting a second lid on to the bottom of a chip box, and then cutting from the second lid and bottom a hole, as shown in Fig. 12 (2); larvæ may then be inserted through the hole; but when the lid is shifted round, and the holes are not opposite, of course there will be no opening, as in Fig. 12 (1), and the contents are secured from escape.

Fig. 12.

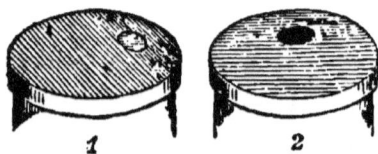

Collecting Box for Larvæ.

Larva preserving is carried out by first killing, and then squeezing and extracting the contents through the anal orifice by means of a crochet hook.

When this has been done, the skin is inflated, but not to such an extent as to distend the segments, and is kept thus inflated while it is being dried in a heated metal chamber. Afterwards, if the colours are observed to have faded, they may be cautiously restored by the application of paint. These objects, mounted on suitable artificial leaves, are then ready for the cabinet.

Chrysalis collecting is conducted according to the situation of the object sought. Some are to be found in the chinks of bark or under loose bark, which may be detached by means of a powerful lever. Some are suspended from trees, bushes, copings, hanging head downwards, or girded by silken threads to low plants or walls; others are to be found in the stems or trunks of their food-plants; many are concealed in cocoons of more or less perfect construction, others again amongst fallen leaves, but the majority are to be met with under the surface of the ground; in which case we shall have to dig for them by the aid of a trowel or broad chisel. The best situations for subterranean pupæ are open park-like fields, borders of streams, open spaces in fir woods, and they are usually situated within a foot or so of the tree trunks, at the depth of two or three inches, though sometimes considerably deeper. Of course both larvæ and pupæ of aquatic species will have to be sought for in their element, among the plants they frequent.

Chrysalis preserving is a simple matter : the pupæ may be killed by plunging them into hot water or by baking; frequently, however, we find that the natural polish disappears with death, and this may be restored by varnishing. It is advisable that the cocoons also, where practicable, should be preserved, to give a notion of their appearance in nature.

E

Moths and butterflies may be sought for at rest or on the wing. They may be disturbed from their hiding places or they may be attracted by various alluring baits.

At rest on stems of grasses and other plants butterflies may be taken on dull, sunless days; but it requires some experience to detect a butterfly with its wings raised up over its back : the little "Blues" may thus be freely boxed in their localities. Again, such butterflies as hybernate may be found in old sheds and outhouses, or under stacks.

Moths may be taken at rest on tree trunks, palings, and walls, or amongst foliage and ground herbage. Some species are to be freely captured in this way after their evening flight is over. Of course, for evening work, a lantern to assist our vision will be indispensable.

On the wing, some butterflies are exceedingly active, others comparatively sluggish ; some fly high, others low. In hunting them, the chief points to be remembered are not to alarm, but rather cautiously to stalk our game, and strike, when we have an opportunity, with precision. It is important also to avoid throwing a shadow over them, and it is a good plan to get to windward of them— anything like flurry will be fatal to success.

Moths which fly by day may be chased in the

same manner, but some may be observed disporting themselves round trees; these must be watched, and netted as they now and then descend. Others fly at a very low altitude, and are only brought into the field of vision by our assumption of the recumbent position. At night again, though we watch for anything stirring the air, among the trees or the herbage, our tactics are somewhat modified; for, if the insect be of whitish colour, we should so place ourselves that its form will stand boldly out against a mass of dark foliage, whereas, if it be dingy in hue, we must take the sky for our background.

Disturbing insects, and thus causing them to start forth, and so render themselves visible, is another method of collecting. This is carried out in various ways.

First, the occupants of high trees may be expelled by jarring the trunk with a heavily loaded mallet, or by thwacking the trunk with a long hazel stick; but a sharp look-out must be kept, for some sham death, and fall plump down, while others make off as fast as they can. Other plans are to pelt the trees with stones, or pump on them with a powerful garden engine, or beat them with a long pole; and of all trees the most profitable for this purpose is the yew; though firs, oaks, beeches, and other trees are not to be despised.

For beating bushes there is nothing better than
a walking-stick, and for low herbage a long switch
passed quickly from side to side with a tapping
movement is best adapted. The tenants of tree
trunks may be disturbed by brushing the surface
with a leafy little bough, or, better still, by the use
of a strong fan, with which a powerful blast may be
driven, the net being held in such a position as to
intercept such insects as are blown off.

Thatch-beating in the autumn is a very profitable
employment, particularly in the matter of *Depres-
sariæ.* Sweeping need only be mentioned here, for
moths collected by the process are anything but
perfect insects.

There are various methods of attracting moths and
butterflies. The first is effected by confining a virgin
female in a muslin cage, the frame
of which may be very readily formed
by bending three pieces of cane into
circles, and fixing these together at
right angles, as shown in Fig. 13.
When this baited cage is placed in a
favourable position, and the weather
is propitious for the flight of the
males, the latter will, in some cases, congregate, and
may be freely captured.

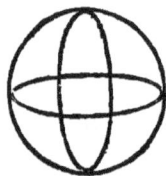

Fig. 13.

Frame of Cage for
Virgin Lepidoptera.

Then, the food-plant of the species is an attraction

at which we stand the best chance of procuring impregnated females.

· Various kinds of blooms possess alluring qualities
for insects: of these, sallow and ivy are the greatest
favourites with collectors. They should be worked
after dusk by means of a lantern and net; but the
combination of a lantern fixed to a long stick, with a

Fig. 14.

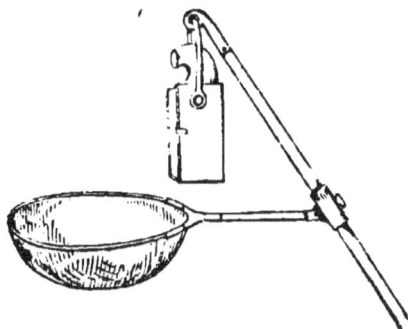

Lantern and Net.

shallow net beneath and a little in advance of it, as
shown in the cut, is the apparatus best adapted for
the purpose; the object of the net being to intercept
any insects which may happen to fall under the
stimulus of light. These attractions should be first
well searched over, and afterwards, a sheet (split if
necessary) having been carefully spread below the
bushes, a gentle shaking should be administered.
Besides these blossoms, heather, ragwort, bugloss,

catchfly, bramble, various grasses, and a vast number of other flowers, are wonderfully attractive. In working patches of bloom we should remain stationary and strike as the visitors arrive. Again,. over-ripe fruit, the juicy buds of certain trees, sap exuding from wounds in trees, are all more or less attractive. The secretion of aphides, commonly called honeydew, observable in hot seasons on the leaves of nettles and various other plants and trees, is also well worth attention, and is at times very productive of insects.

Sugaring is the next attraction, and a very important one it is. "Sugar" may be prepared by boiling up equal quantities of coarse "foots" sugar and treacle in a sufficient quantity of stale beer, a small quantity of rum being added previous to use, and also, if considered advisable, a flavouring of jargonelle pears, anise-seed, or ginger-grass. This mixture should be applied by means of a small paint brush to the trunks of trees, to foliage, flowers, tufts of grass, or indeed to any object which may present a suitable surface; for in some localities we are put to shift to know where to spread our sweets. This operation should be performed just before dusk, and soon afterwards the baited spots should be visited and, by the aid of a lantern gently turned on them, examined, a net being held beneath the while. The best form of net for the purpose is formed by socketing two paragon wires into a Y-piece and connecting their diverging extremities by a piece of catgut, as shown in Fig. 15. The catgut, being flexible, will adapt itself (see the dotted line) to the surface of a tree trunk when pressed against it. With regard to insects captured at sugar, they are usually remark-

Fig. 15.

Fig. 16.

Box with linen joints.

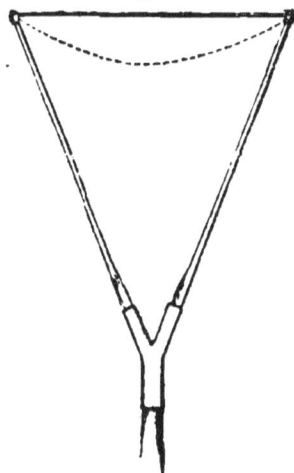

Net for sugaring.

ably quiet, and may be boxed without difficulty, and, with a few exceptions, may be conveyed home in the boxes, care being taken to let each have a separate apartment. The boxes should be strengthened with strips of linen pasted round the joints, as shown in Fig. 16, otherwise accidents may occur, particularly on wet evenings or on rough ground. The skittish

individuals may be best captured by means of the
sugaring drum, of which a cut is given in Fig. 17.
This apparatus consists of a cylinder, one end of
which is covered with gauze, the other provided
with a circular valve, which works in a slit. For
use, the valve is opened and the cylinder placed over

Fig 17

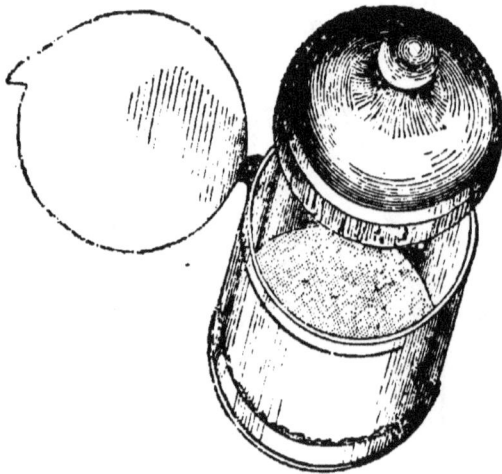

Sugaring Drum.

the insect, which naturally flies towards the gauze ;
then the valve is closed, the corked piston, shown at
the upper part of the cut, placed against it, the valve
re-opened, the piston pushed up to the gauze, the
insect pinned through the gauze, and the piston
withdrawn with the insect transfixed to it.

Light is another most profitable means of attracting. The simplest way is to place a powerful mode-rator lamp upon a table in front of an open window which faces a good locality, and then wait net in hand for our visitors, which usually make their appearance late in the evening, and continue to arrive until the small hours. Those who prefer it can use the American moth-trap, which is self-acting, detaining such insects as may enter its portals, or those who can afford the space may fit up a room on the same principle. Street lamps are very profitable certain localities, and amply reward the collector who perseveringly and minutely

Fig. 18

Cyanide Bottle and Ferrule.

examines them. The apparatus depicted in Fig. 18 is very useful for taking off such insects as may be on the glass of the lamp: it consists of a cyanide bottle attached by a ferrule to the end of a suffi-ciently long stick. When placed over an insect, stupefaction is quickly produced. A net of the shape represented in Fig. 19 is also very useful for getting at the various parts of the lamp.

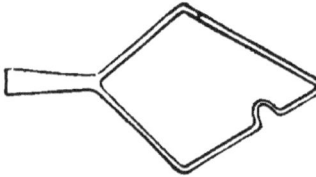

Fig. 19.

Lamp Net.

The best methods of stupefying and killing insects on the field is the cyanide bottle, prepared by placing alternate layers of cyanide of potassium and blotting-paper in the bottom of a wide-mouthed bottle, the mouth of which is accurately stopped with a cover, which is better for the purpose than a bung. The chloroform bottle, which is generally made with a little nipple, through which the fluid flows slowly out, and covered with a screw-tap, as in the cut 20, is also handy. The chloroform should be dropped over perforations in the box containing our patient, these perforations having been previously made by a few stabs of a penknife. After the fluid is dropped, our thumb should cover it, when the vapour will quickly enter, and the inmate speedily become insensible.

Fig. 20.

Chloroform Bottle.

Afterwards the *coup de grâce* may be given to the insect by pricking it under the thorax with the nib of a steel pen dipped in a saturated solution of oxalic acid. If we are smokers, a puff of tobacco may be blown into the box with like result. If we are destitute of any apparatus, and brimstone lucifers for the purpose of suffocating our captures under an inverted tumbler cannot be obtained at some roadside inn, we must fall back on the barbarous practice of pinching the thoraces of such as cannot be carried home in boxes. At home we shall find the laurel jar and ammonia bottle the most useful. The former is made by partially filling a large wide-mouthed bottle or jar with cut and bruised dry leaves of young laurel : if any dampness hang about them, we shall have the mortification of seeing our specimens become mildewed. The latter consists in adding a few lumps of carbonate of ammonia, or some drops of strong liquid ammonia, on a sponge, to the bottle in which our captures, with each box lid slightly opened, have been placed. But it must be borne well in mind, firstly, that ammonia is injurious to the colours of most green insects; and secondly, that if the specimens be not well aired after having been thus killed, the pins with which they are transfixed will become brittle and break. Insects should be left in the ammonia for several hours, and are then in the most delightful condition for setting out.

To pin an insect properly is a most important

procedure. The moth, if of moderate dimensions may be rested or held between the thumb and fore finger of the left hand, while the corresponding digits of the right hand operate by steadily pushing a pin through the thorax, bringing it out between the hind pair of coxæ until sufficient of the pin is exposed beneath to steady the insect in the cabinet. The direction of the pin should be perpendicular when the insect is viewed from the front, as in

Fig. 21   Fig. 22.

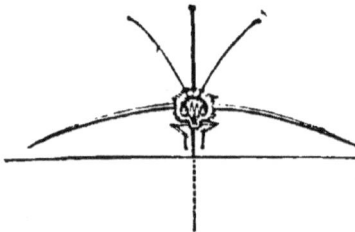

Front View of properly pinned   Side View of ditto.
        Insect.

Fig. 21; but a lateral view should show the pin slightly slanting forwards, as in Fig. 22. Pins made for the purpose in numerous sizes are sold by Mr. Cooke, of New Oxford Street. '

Setting out moths and butterflies is an operation which, if skilfully performed, adds much to the beauty of the future specimens. The method of setting most popular is carried out by means of saddles and braces. These so-called saddles are pieces of cork rounded as in the sectional figure, a groove being cut out for the reception of the bodies of the insects: they are generally strengthened by

Fig. 23.

Cork Saddle for setting out Insects.

a strip of wood, upon which they are glued. Braces are wedge-shaped pieces of card or thick note-paper, the thick end strengthened, if necessary, with a disc of card fixed by shoe-makers' paste, and pierced with a pin through it, as shown in Fig. 24. The mode of application of these appliances is beau-tifully shown in Fig. 26.* But before these straps can be applied, the wings must first be got into position by means of the setting needle and setting bristle, which are thus manipulated (the set-ting bristle, by the way, being formed by fixing a cat's whisker and a pin into a piece of cork, at

Fig. 24.

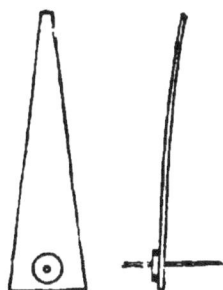

Braces for setting out.

---

* Figs. 26 and 27 have been kindly lent by Messrs. Reeve & Co.

the angle shown in Fig. 25):—After the insect is
straightly pinned upon the saddle, and the legs,

Fig. 25.

Setting Bristle.

antennæ, and, if necessary, the tongue, got into
position, the left fore-wing is to be pushed or tilted
into its place by means of the setting needle, which

Fig. 26.

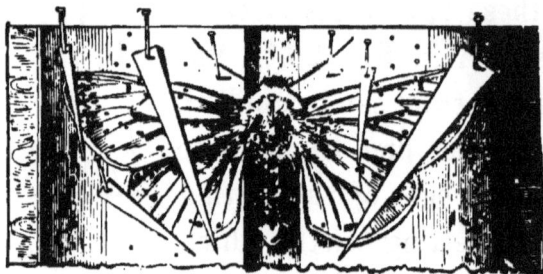

Moth set out on Cork Saddle.

is merely a darning needle with a handle; and simultaneously it is to be held down by the bristle; then
a small brace should be applied to the costa of

the fore wing. Next the hind wing should in like manner be adjusted, and as many braces as are considered necessary to keep the wings in their place should be added. Lastly, the right side of the insect should be treated in a similar way.

A very useful mode of setting, invaluable when we are destitute of saddles, is known as "four-strap" setting, and is well explained in Fig. 27. In this

Fig. 27.

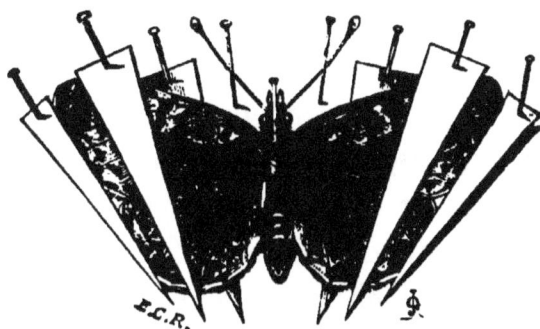

Four-strap setting.

case the lower straps are first put into such a position, that when the insect is placed over them the middle of each of the costæ will rest upon them; then the wings are got into position, and the second pair of straps are applied over the wings, the latter retaining their position through the elasticity of their costæ: two more straps are generally added to secure the outer borders of the wings, as shown in the drawing; but these, though advantageous, are not absolutely necessary. The saddles, with their contents, should be kept in a drying house, which is a box adapted for their reception, and freely ventilated, until the specimens are thoroughly dry, when the latter may be cautiously removed, and transferred to the collection.

To preserve our collection from decay, considerable care and attention is necessary. In the first place no insect which is in the least degree suspected of being affected by mites, or mould, or grease, should on any account be admitted to our collections. It is best to be on the safe side and submit every insect received from correspondents, whether mity or not, to quarantine, by which is meant their detention for a few weeks in a box the atmosphere of which is impregnated with some vapour destructive to insect life; such as that of benzole. Our own specimens we should kyanize by touching the bodies of each with a camel's-hair brush dipped in a solution of bichloride of mercury of the strength six grains to the ounce of spirits of wine,—no stronger.

As for mould, it is best destroyed by the application of phænic or carbolic acid, mixed with three parts of ether or spirit. As preventives, the specimens should be kyanized as above. Caution in the use of laurel as a killing agent must be exercised, and the collection must be kept in a dry room.

Grease may be removed by soaking the insects in pure rectified naphtha or benzole, even by boiling them in it if necessary. When the bodies only are greasy, they may be broken off, numbered, and treated as above. After the grease is thoroughly softened, the insects should be covered up in powdered pipeclay or French chalk, which may be subsequently removed by means of a small sable brush. As a precaution against grease, it is advisable to remove the contents of the abdomina by slitting up the latter beneath with a finely pointed pair of scissors before they are thoroughly dry, and packing the cavities with cotton wool. The males, especially of such species as have internal feeding larvæ, should be thus treated.

Some prefer to keep their collections in well-made store boxes, which possess many advantages over the cabinet ; for example, they may be kept like books in a bookcase, the upright position rendering the contents less liable to the attacks of mites; they are more readily referred to, and are more portable, and they admit of our gradually expanding our collections to any extent. Cabinets, on the other hand, are preferred by many, for the reasons that they are compact and generally form a handsome article

F

of furniture ; moreover, good cabinets are made entirely of mahogany, which is the best wood for the purpose ; deal, and other woods containing resinous matter, having a decidedly injurious effect on the specimens. As a preservative, there is, after all, perhaps nothing better than camphor; but it should be used sparingly, or its tendency will be to cause greasiness of the specimens.

V.

BEETLES.

By E. C. Rye, F.Z.S., etc.

The general rules, so ably expounded by Dr. Knaggs in his instructions for collecting *Lepidoptera*, as to constant alertness and making "the reason why" the starting-point of investigation, apply with equal force to the collector of *Coleoptera*, and need not be here recapitulated. But they do not, in the instance of the latter, require generally to be observed, except as to the perfect state of beetles; for, owing to the hidden earlier conditions of life of most of those insects, and to the long period during which these conditions exist, it is but seldom that the pursuit of rearing them, so universally and profitably adopted by the Lepidopterist, is found of much use to the collector of beetles. And this is very much to be regretted; because, in the majority of cases, if the latter succeed in rearing a beetle from its earliest stage, and keep proper notes of its appearance and habits, he will probably be adding to the general stock of knowledge, as the lives of comparatively few, even of the commonest species, are recorded from the beginning. It may be, also, in addition

F 2

to the reasons above mentioned, for the usual want of success attending the rearing of beetle larvæ, that the fact of bred specimens being frequently (from the artificial conditions attending their development, and from their not being allowed that length of time which, in a state ot nature, they require after their final change before they are ready to take an active part in their last stage of life) not nearly so good as those taken at large, militates considerably against the more general use of this method of adding to a collection. In this respect, of course, the Lepidopterist is actuated by precisely opposite motives; as tor him, a bred specimen is immeasurably superior to one captured. And the fact of so few beetle larvæ being known at all, or, if known, only to the possessor of somewhat rare books, renders it very likely that a mere collector, finding a considerable expenditure of patience and trouble result in the rearing ot a species of which he could at any time readily procure any number of specimens, may very probably abandon rearing for the future.

These observations, however, are not in the least intended to dissuade anyone from breeding or endeavouring to breed beetles. On the contrary, it is obvious from them that it is precisely by attending to these earlier stages that the earnest student (novice or expert) has the most chance of distinguishing himself, on account of the more open field for discovery. And in the instance of many small, and especially gregarious, beetles, breeding from the larvæ is frequently very easy, if only the substances (fungus, rotten wood, roots, stems of plants, &c.) containing them be carefully left in precisely the state as when found, and be exposed to the same atmospheric or other important conditions. In fact, to ensure success and good specimens, it is best that in their early stages beetles should be "let alone severely."

It may be here observed that we have been lately in this country indebted to the minute observations and great tact of some of our best students of *Micro-Lepidoptera* (in which branch of entomology we are second to none in Europe) for some most interesting additions to our knowledge of habits, and for long series of beetles usually rare in collections.

Dismissing then the earlier stages of beetles, the following observations will apply only to the imago, or "beetle proper." And here I would repeat how evident it is that the knowledge of "the reason why" is especially indispensable to the beetle collector, judging from the extreme rarity of the occurrence of any new or valuable insect in the stores of a mere random collector or a beginner. For him, no old hand detects an equivalent to *Daplidice* or *La-*

thonia in his duplicate boxes ; whereas, among Lepidopterists, " school-boy's luck " is proverbial. I can
give no reason for this statement, founded on my
own (by no means trifling) experience in the way of
examining specimens. And in this idea I think I
am corroborated by the very great rarity in old
collections and records of many species now universally common; the directions in older manuals,
as to looking under stones, on walls, paths, &c.,
pretty clearly showing that the majority of captures
in the olden time were what are now irreverently
designed as "flukes." Still, it is astonishing to what
good account a sharp observer may turn these casual
meetings, often to him resulting in the discovery of
"the reason why " as to the particular species acci-
dentally found, and to the correlative increase of his
collection. And, apart from captures during collecting expeditions, good things will at times occur to
the alert entomologist: one, for instance, who will
startle his friends in the streets by suddenly swoop-
ing with his hat after an atom flying in the sunshine,
or who is not too proud to pick up another, racing on
the hot pavement, during those days of early spring,
when the insect myriads, revelling in warmth and
light, after their long winter's durance, may be seen
madly dashing about, even in towns: on such a day,
for instance, as that whereon a certain well-known

doctor among the beetles found that living carabi-
deous gem *Anchomenus sexpunctatus*, far from its
native *Sphagnum* and heath, wandering on the flag-
stones of the W.C. district.

But, before referring to special modes of hunting,
it may be as well to mention the *instrumenta belli*
required for the equipment of the Coleopterist in
this country. These are but few, and of the simplest
kind; indeed, in entomology, as in the gentle art of
angling, it is often the most roughly accoutred that
secures the best basket. The umbrella net, figured
at p. 47, used both for beating into and sweeping,
cannot be dispensed with, and a beating stick can
be cut out of the nearest hedge. The net itself
should be of fine "cheese cloth," or some strong
fabric that allows the passage of air, but not of
beetles; otherwise, if of too close a fibre, it is apt
to "bag" with the inclosed air, and reject its contents during the operation of sweeping. The net
being of course used with the right hand, its left
top edge especially bears the brunt of the attendant
friction, and gets soon worn; it is consequently
advisable to have an outer strip of stout "leather-
cloth" sewn strongly over the rim there for some
little distance, extending that protection also to the
right top edge, though not for so long a space. The
curved handle of the stick should be sawn off as soon

as possible; it frequently catches in the pockets of the sweeper, causing a jerk to the net, and dispersal of its contents. For a similar reason, the ferruled apex may well be removed. Some collectors keep the sharp cutting edge of the spring sides of the net uncovered, sewing the net itself to holes drilled at intervals on the lower side of the springs: a net of this kind cuts very close, and where there is much herbage soon gets full of fragments, taking a long time to examine. It will be found handy if the bag of the net be cut to a point from the front towards the handle side: this causes the contents to gravitate to the bottom, as far as possible from the point where the rim meets the substance swept.

A common umbrella (easily slung by a stout string over the back when not in use) is an admirable (some think, superior) substitute for this net, as it can be held up higher by the ferrule, and tall bushes and trees (of which the branches nearer the top are usually most productive) can be beaten into it with more certainty of their beetle contents being caught. The steel frames will be found in the way when the beetles are being bottled; consequently, a good large gingham may be consecrated to collecting, and its inside (not merely the outer ribs) covered all over up to the middle (leaving no aperture there if possible) with thin white calico, stitched *over* the frame.

Another good form of net for sweeping or dragging in long grass or herbage, is of the common fishing landing-net description, made of very stout wrought-iron or steel wire, either in a simple hoop, if a moderate size only be required, or with a single hinge to fold into two, or with three such hinges, folding into four, as may be desired. I have used one of these four-folding nets for, years, and never found it fail. One end is hammered out flat and perforated, the other forming a male screw (1½ inch long), bent at right angles with the body of the frame, passes through the hole, and fits into a female screw in a strong and long ferrule, fixed in the usual way to the end of a stout oaken walking-stick. As the power exerted in sweeping with such a net is great, and the action continuous, the simple screw is not enough, and a small screw hole is drilled right through the ferrule and the screw end of the net; a small thumb-screw, in shape like an old-fashioned clock key, going transversely through both, and effectually hindering lateral displacement. The framework of the net and the ferrule are better made of the same metal, because, if made of two metals of different density, the stronger soon wears away the weaker; and the stick must be inserted deeply into the ferrule, and held on with *two* deep pins or small screws on opposite sides (not on the same level, however, as the wood is in that case weakened), one being insufficient to stand the strain. The net, of the same substance as that above mentioned, is made with a loose "hem" to slip on the frame before screwing it in the ferrule. A leather-cloth edging all round is advisable, and the bag should be cut long enough to prevent the possibility of the contents jerking out. Another very good plan for securing the frame to the ferrule is to have both ends of it soldered together into a deep square-sided plug, fitting into a corresponding square hole in the ferrule. The small cross-screw or pin is here also to be used; but the angles of the plug naturally keep a much tighter hold than the worm of a screw. Such a frame as this cannot, of course, be folded.

For water beetles, a similar net to that last mentioned is effective, but it should be stouter and with a flat front, for dredging closely against the sides and bottoms of ponds. The best substance for its bag is fine sampler canvas; and a very large, stout bamboo cane is at once light and strong for its stick. To avoid friction the bag may be affixed to small wire rings let into holes on the lower edge of the frame, or running on the frame itself.

A sieve is one of the most remunerative implements, and may be procured either simple or folding. It consists of a stout wire-framed circle, connected by a strong linen band, six inches deep, with the bottom of an ordinary wire sieve, the meshes of which are wide enough to allow any beetle to pass through. Leaves, grass, flood refuse, ants' nests materials, cut grass, seaweed, haystack and other *débris*, are roughly shaken in this over a sheet of brown paper, which should invariably form part of a Coleopterist's apparatus. A stout piece of double waterproof material may be substituted ; and, in marsh collecting, must be used as a kneeling pad.

For ordinary bark collecting, a strong ripping chisel (of which the blade is well collared, so as not to slip) is as useful a tool as can be procured; but for *real* tree working, no ordinarily portable instrument is thoroughly effective. Light steel hammers with a lever spike may delude the collector ; but a woodman's axe, a saw, a pickaxe or crowbar, will often be found not too strong. For cutting tufts, &c., a strong garden pruning-knife is good, and an old fixed-handled dinner-knife (carried in a sheath) better. For holding the results of the operation of these instruments, the collector needs but one or two collecting bottles—one rather small and circular, of as clear and strong glass (*not* cast) as can be got, with a wide mouth and flat bottom. Its neck should not slope, but be of even width, or the cork will often get out of itself. This cork should be a deep one, and be perforated longitudinally by a stout and large round quill, the bottom of which should be level with the bottom of the cork, the top projecting some inch and a half, with the upper orifice not cut off straight, but slightly sloped diagonally, so as more easily to scoop up beetles from the net or hand.　It is closed with an *accurately fitting*, soft, wooden plug, rather longer than the quill, reaching exactly to the bottom of it, but with its top projecting above the top of the quill, and broader than it, so as to be easily pulled out by the teeth when the hands are occupied.　The bottle should be secured by stout twine to the buttonhole, enough play being left for it to reach the net in any ordinary position.　I usually secure the external junction of quill and cork with red sealing-wax, and have more than once found the bright red catch my eye when I have lost my bottle. [N.B. This loss will always happen to every collector; generally after a peculiarly lucky day's work: so use the string-preventive.]　The body of the bottle may usefully be half covered with white paper gummed on.　A few stout, plain glass tubes, papered in like way, and with plain corks, should be carried for special captures; and a cyanide bottle,* as mentioned

---

* "Killing bottles," containing cyanide of potassium under a layer of gypsum, may be bought at most natural-history apparatus dealers, and are useful as relaxing dépôts.

at p. 57, or one containing bruised and shredded
young laurel shoots, will be found useful for safely
bringing home larger species, or such as would
devour their fellow-captives. When put into these,
beetles almost instantly die and become rigid, needing
a stay of two days or so to become relaxed, in which
condition they will then safely remain for a considerable period. In the first collecting bottle a piece of
muslin should be put, to give the contents foothold :
these are brought home alive, and killed by bodily immersion in boiling water, after which they are placed
on blotting paper to drain off superfluous moisture.

Good things should always, when practicable, be
set out at once, as the pubescence is apt to get matted
if they are consigned for too long a period to the
laurel or cyanide bottle ; but such as remain un-
mounted can be put in a little muslin bag, and deposited in laurel until a more convenient opportunity.
Beetles also, when taken in large numbers during an
expedition into a productive locality, may be collected
indiscriminately into a bottle containing sawdust
(sifted to get rid both of large pieces and actual
dust), slightly alcoholized, or with a small quantity
of carbolic acid or cyanide of potassium in it. Each
night, on reaching home, these will be found to be
dead, and they can then be transferred to a larger
bottle or air-tight tin can, partially filled with the

same materials and a little carbolic acid to check undue moisture. Filled up with sawdust, this will travel in safety for any distance, and almost any time.

Species of moderate size, say up to that of an ordinary *Harpalus*, are in this country usually mounted on card. Much is to be said both for and against this practice: it enables the proportions and formation of limbs to be well appreciated, and it preserves the specimens securely; but there can be no doubt that it prevents an inspection of the under side, except at the slight trouble of extra manipulation in floating off in cold water and reversing, and that the gum used clogs the smaller portions of the insect that come in contact with it. Specimens larger than those mentioned should be pinned through the centre of the upper third of the right wing-case (never through the scutellum or thorax), and the limbs extended in position with pins on a setting board, made of a flat strip of cork glued on deal. Both these and the mounted examples must be left to dry, for a week at least, in the open air: if the boards are fitted in a frame, they can be reversed (as soon as the gum is dry in mounted specimens), so that the specimens are bottom upwards—dust cannot then collect on them, and there is less chance of mites attacking them. Specimens dry more rapidly in spring and summer than at any other time, and of course more readily in dry weather.

For mounting specimens, five or six small pieces of the finest and most transparent gum tragacanth, or "gum dragon," with rather less than the same number of pieces of clear gum arabic, are to be put in a wide-mouthed bottle with about a large wine-glassful of cold water. In a short time (twenty-four hours at most) the gum absorbs the fluid and swells; then add half as much more water, and stir the mixture, which, on being left for another twenty-four hours at most, will be ready for use. The mixture should be dull white, of even texture, and not quite fluid. Never make a large quantity at one time, or be persuaded to put *anything* else into it. Card for mounting should be the whitest, smoothest, and best that can be procured. "Four-sheet Bristol board" for large specimens, and three-sheet for ordinary use, are about the proper degrees of thickness. Robersons, of Long Acre, artists' colourmen, have promised the writer to turn out cardboard of this kind with an extra milling, to ensure a good surface. Upon strips of this card, pinned on a setting board, the insects to be set out are mounted, one at a time, and not too close to each other, each on a separate "dab" of the gum, the limbs being duly set out with a fine pin or needle mounted in a paint-brush stick. A pin with the point very finely turned, so as to form a minute hook, is very useful ; and for extremely minute work a " bead-needle " is valuable. The gum-brush should not be used in setting, but one or two very fine-pointed camel's-hair brushes may be found of advantage. Before mounting, reverse the specimen on the blotting-paper, and brush out its limbs as far as practicable with a damp flat brush. Very refractory individuals may require to be gummed on their backs ; as soon as the gum is dry, their limbs can be more easily got into position, and they can then be gently damped off their temporary mount, and treated as above.

A small pair of brass microscope-forceps, ground or cut to a minute point, will often materially assist in getting refractory limbs into position. French white liquid glue (not made of shell-lac) is useful for fastening down larger specimens, as it is very strong and dries readily ; and with a very small quantity of it rows of specimens can quickly and securely be roughly mounted, in the Continental way, which is preferable in many cases to leaving the insects for a long time in laurel before setting them out. Such specimens can afterwards, if desired, be relaxed by leaving them on damp sand, or in the cyanide or laurel bottle, and be then set in the way above indicated.

Care must be taken, in setting, not to put the specimen lop-sided on the card, or to distort its segments unnaturally by pulling them out of position, &c., and not to allow gum to lodge anywhere on the upper surface. It is easy, soon after a specimen is securely mounted, to remove with clean water and brush any superfluous gum. In preparing such insects as are liable to "run up" in drying (e. g. the *Staphylinidæ*), the abdomen should be duly pulled out by a bead-needle inserted at its apex; and to prevent the contraction of the internal muscles in drying, this part may be held with the liquid glue above mentioned. Usually, by putting these insects as soon as mounted into a box and keeping it closed for a few hours, while the first drying takes place, the proper dimensions of the abdomen may be preserved, and thus the natural facies of the insect retained. The contents of the bodies of very large insects may well be removed, either by the anal orifice, or by an incision on the lower side of the abdomen. The Oil-beetles (*Meloë*) alone require careful stuffing. This is best done by separating the entire abdomen from the metathorax, beneath the elytra, and close to their point of insertion: the body is then easily emptied and washed out, and may be filled with cut-up wool, which packs closely; when gummed on again, the junction is not visible, and the entire insect preserves its wonderfully obese appearance.

To save time, in mounting many specimens, it is better to merely gum straight on the strip of card as many specimens as can be managed at a sitting. The left side of each of these can then be slightly damped with clear cold water, and its left limbs set out: when all are thus done, the first one will be nearly, if not quite, ready to have its right side treated in like manner; and so on to the end. Very refractory specimens will sometimes require to be even held down with little braces of card on pins, and to have each limb damped and set out by a separate operation. The card of large specimens will often curl upwards in drying, owing to the amount of damp: to counteract this, the *lower* face of the card may be washed with a wet brush, just before gumming its surface.

Before putting insects away, when dry, the individual specimens should be cut off the strips of card by a straight cut on each side, one at right angles to the sides in front, and another behind, all (except the last) close to the tips of the limbs as set out, so that the whole card forms a parallelogram. A very little practice will enable the operator to do this both certainly and quickly. No two individuals (save perhaps a male and female, of whose sexual relations there can be no doubt, or an example mounted on its back, to show its under side, along with a member of the same species) should be allowed to continue on one card; much less should *a row* be left together. The reason of this is, that in many cases species closely resembling each other often get confused; and it is, moreover, difficult to get a glass of anything but a very low power to bear upon all parts of the individuals without injuring some of them. Each specimen should have sufficient card left *behind* it to allow of a glass of high power being passed between it everywhere and its pin. The pin should perforate the card in the middle of, and close to, its hinder margin; and the whole card be lifted three-fourths up the pin, to keep it from mites and dirt as much as possible. Proper entomological pins can be obtained of all sizes at the agents of Edelsten, 17, Silver Street, St. Martin's-le-Grand; also (with all other apparatus) of any natural-history agent or dealer in London; such as Mr. E. W. Janson, 28, Museum Street, or Cooke, New Oxford Street. "No. 8" pin is, perhaps, the most useful size. In removing many specimens, proper insect forceps will be found handy : these can be obtained at the two last addresses; or of Buck, cutler, Tottenham Court Road.

Specimens will occasionally become discoloured with grease, usually from defective drying, though many water beetles and internal feeders, and most autumn-caught specimens, are specially liable to this defect. Benzine is an effectual remedy for it and for mites, and can be liberally applied with a brush. Carbolic or phænic acid, dissolved in that fluid (or alone, see p. 64), is an effectual safeguard against mould from damp; and when in solution with water, this acid has been found useful as a wash for card and boxes, which then are not attacked by mites. To re-card a specimen that has become discoloured (whether from either of these causes, or from age), it is only necessary that it should be floated in cold water for a few 'minutes; the insect can then be dried, well saturated with benzine, and again mounted, looking as fresh as ever. But, in re-carding specimens, it is necessary to be very careful with such as were originally kept too long in a laurel or cyanide bottle, as they are apt to become so rotten that a little damp will cause a "solution of continuity."

As to storing the specimens when quite dry, I can add nothing to the excellent observations of Dr. Knaggs, at p. 65; the same remarks applying with equal force to *Coleoptera*; except, perhaps, that, even when the collector has (and is satisfied with) a cabinet, he is likely, in proportion to the *real* work done by him, to establish type-boxes of all the difficult groups.

For the examination of insects, readily manipulated by being pinned singly on a square, flat, thick
piece of cork or bung, a pocket glass is, of course,
necessary. In this case, the best instrument is the
cheapest in the long run, whatever its cost; and one
by a good maker, such as Ross, with modifications of
four powers, will suffice for any ordinary work. For
very small species, a Coddington, of the clearest
definition and highest power attainable, is of im-
mense help. But when the collector finds that he
needs a compound microscope to separate species,
it is the firm opinion of the present writer that
that collector had better take to some other pursuit
than studying *Coleoptera*. To anyone, however, whose
researches entail an examination of the minute
cibarian and other organs of beetles, whether for
purposes of classification or otherwise, the compound
is absolutely necessary; though even then the lower
powers are usually sufficient. For rough dissection,
all that is needed are an oculist's very small lance-
headed dissecting-knife and a stout and fine needle.
With these, under a lens mounted on a little stage
to allow the free use of both hands, much may be
done. The writer, however, has seen and used a very
pretty (and comparatively inexpensive) dissecting-
stand, with various powers and much latitude of
motion, by Ross.

After mentioning that, in sending mounted beetles by post to correspondents, it is far more practical to use a *strong* box, not too deep, to fasten the pins securely, with a layer of manufactured wool in the lid (glazed side towards the beetles, so as not to catch limbs), and to put more wool outside, and write the address and affix the stamp on a label attached, than it is to pack carelessly, write "With care" outside,* and then grumble at the post-office because the insects are broken,—I think I cannot, with use, say anything more upon beetles in their preserved condition; and I will therefore now give some hints as to their haunts when alive.

To exhaust the accidental-capture system above alluded to, mention must first be made of sand-pit collecting, a most profitable employment, especially in spring and early summer. A clean, straight-sided silver-sand pit is the best, and if in or near a wood its attractions will be at their highest. Beetles, flying of an evening and by night, dash against the pit sides and fall to the bottom; others merely crawl in for shelter, or tumble over the sides, and many seem attracted by the mere damp at the bottom or in the corners. Old collectors used to recommend a

* It is, however, always as well to write "Insects," signifying contents that are "caviare to the million," and therefore not likely to be appropriated *en route.*

sheet spread out to attract insects; and there is no doubt that a certain number can be found by such means, just as they can be picked up floating on horse-troughs or on ponds. Artificial traps exist in the corridors of the Crystal Palace, some half-inclosed country railway stations, and such places; crawling up the windows of which many specimens are to be found. But these can only be considered as indications of what species occur in the district, as they are mere stragglers. Deliberately laying traps in sand pits, on commons, &c., will be found most productive. Small dead animals, fir branches, dead leaves, &c., can be examined time after time with profit in such situations. Burying a stout branch with the bark on, leaving the top above the soil, and periodically examining it when damp and nearly rotten, has been found effective; many insects collecting beneath the loose wet bark.

After heavy floods, as during severe droughts, beetles may be found in great profusion; in the former case, by sifting the refuse left by the water; in the latter, by diligently examining the damp residuum of former ponds, and if no damp be found, by even searching below the surface where it last occurred.

The wet hay, often decayed and mouldy, at the bottoms of stacks, which bad farmers have placed directly on the ground, will be found to teem with beetle-life; as will the margins of dung and vege-table-refuse heaps, wood-stacks, cut grass, &c.; and many good things may be taken by gently waving a light gauze net to and fro, just before sunset, close to such places, whither the instinct of nature impels the flight of myriads.

In winter, isolated tufts of grass in wet places, on the margins of streams, the crests of banks, &c., must be cut close to the ground, and gently torn in pieces over brown paper. Wherever many insects seem to be found, it will in most of these cases be found advisable to sift the fragments, and bring home the beetles and small stuff unexamined in a bag with a string at the neck to prevent their escape. Moss should be treated in this way, and the layers of black and rotting leaves found in woods, especially at their outskirts. Beech leaves usually produce many species, and the autumn and spring are the best times for hunting for them.

In winter, also, many species will be found hybernating in grass at the roots of trees, under bark, &c., in conditions not usual with them at other times.

In autumn, fungi, in woods especially, will be found most productive.

General sweeping, except during the winter, will

always be more or less remunerative. No general
rules can be laid down for this; in a good neighbourhood (on chalk or sand, or, better still, in a district
where both these soils are found) beetles will swarm
almost anywhere in due season, and the most un-
likely-looking spots will frequently be found the
best in the end. In luxuriant herbage, among low
shrubs, in the close-growing vegetation of hill-sides,
the sweeping net may be plied with success; but the
best way, with all *Phytophaga* at least, is to start
with a fixed idea as to catching certain definite
species, and then, at the right time, to hunt for such
plants as these are known or supposed to frequent;
and, such failing in the district, to try their allies.
Of course, the collector will not fail to sweep flowers
in woods and lanes, whereon, in the hot sunshine,
many showy beetles bask. Many good things will
be found by sweeping under fir trees, especially
towards evening, and even by night; in many places,
especially marshes, nocturnal feeders may be secured
by the vague use of this net. By night, also, many
species may be found at sugar put on trees for
moths, and on ivy or sallow-blossom.

Beating is most productive in early summer,
especially in the second year's growth of young
cuttings in woods; and the oak, hazel, and poplar
will generally yield many species to the tap of the

stick. Good thick, and especially *old* hedges, must
also be always carefully thrashed into the net;
very many good things, otherwise not procurable,
will reward this toil. Another scheme for getting
rare species is to beat the tops of trees with a long
pole, placing beneath a sheet or tent covering.

Breaking away the extreme edges of banks,
throwing water on them, treading heavily on the
margins, diligently examining grass and roots close
to the water, reeds (especially if cut and on the
ground in heaps), &c., will bring to light great
numbers of wet-loving beetles. Water beetles,
pure and simple, must be dragged and dredged for,
especially round water plants beneath the surface,
and along the sides of ponds, in eddies of running
streams, in the moss on stones in them, and on the
stones themselves, &c.

The *Coprophaga* will be found readily in the
droppings of various *Mammalia*, and also in holes
bored in the ground beneath, often to a great depth.
An easy and clean way to secure them is to throw
droppings, ground and all, into water, the beetles
coming to the surface.

As to wood beetles, they must be sought for
under and in bark, in solid wood, in decaying
branches, and such places; a rule to be remembered is, that most of these occur at the *tops* of

trees: hence the paucity of so many species in collections. Indeed, to properly hunt for the majority
of them, it is necessary to obtain *carte blanche* and a
ladder, if any success be hoped for. Felled trunks
are, of course, easy to manipulate; and their freshly-
cut stumps, exuding either resin or a peculiar and
often sweet *mucor*, are very attractive to many
beetles, as is freshly-cut sawdust, and most especially the (to us) fetid and acrid juice resulting
from the attacks of the larva of the Goat-moth.
Rotten fruit, &c., are also not to be passed by without examination. Many small species occur in, or
can be obtained from, the topmost twigs of trees
blown down by the wind.

Dead animals, as before mentioned, must be examined, as must the vegetation and soil near them.
A keeper's tree in a wood will always produce something for the collector, who need only hold his net
beneath the gibbeted *feræ* and bang their hides and
bones with his beating-stick. During different
stages of decomposition and desiccation, beetles of
widely varied affinities will result from this method
of collecting.

Ants' nests would require a special notice, so productive are they: their material can be sifted and
their neighbouring "runs" or paths examined,
traps laid near or on them, and periodically cleared

out, &c. Bees' and wasps' nests also produce good, though fewer species, and are, moreover, not quite so easy of access. The nests of birds, especially if the latter are gregarious, and, indeed, the habitations of any animals, will be found to harbour many beetles, amongst other insects.

In gardens, the beetle collector should lay cunning traps of cut grass, twigs, planks, bones, &c.; by a periodical examination of which he will secure many good things. If there be a hothouse about the premises, it and its belongings will always act as a bait.

Large tracts of waste land and commons, though superficially apparently unproductive, often contain congregations of good species, in some little oasis of damp or vegetation; moreover, on them several peculiar beetles occur. Hills and mountains will often suddenly repay the toil of the collector, who has despondently worked his way up, turning over stones, and finding comparatively nothing. The moss, &c., attending the channels of any streams in such places should be carefully searched, and the stones on the top especially not neglected. River banks and salt marshes are invariably frequented by good insects, and the very heaps of seaweed, dry or wet, on the shore harbour countless beetles. In such places small sand-loving plants should be pulled up by the roots, and, with the neighbouring soil, shaken over brown paper. The sand itself may in many instances be scraped, and burrowing beetles brought to light; but if the hunter comes upon a dead fish or bird, a full bottle will be his.

Thus it will be seen that almost every locality contains beetles, if the collector can only detect them (and it may be as well here to impress on him that it is better to bottle a dubious insect and examine it at home, than to reject it for being *apparently* common). Still there can be no doubt that certain soils and districts are much more productive than others; for instance, most of the midland and western counties, and some of the south-western, are not by any means so prolific as the eastern, southern, and many parts of the northern districts of Great Britain; clay being the worst of all soils for the Coleopterist.

The collector will do well, after a first hurried "burst" at all beetles that come in his way, to select a special group, and lay himself out to work it carefully, buying or borrowing the works of authorities upon it, and making himself master of the *botany* connected with it, if it be a group of plant-frequenting habits. By such a way of working, he will more quickly, though step by step, acquire a good collection, and a stock of useful knowledge, than by any other. He will of course keep a register of the date and place of capture, and any peculiarity of habit of each insect he takes. Figures of the date of the year (usually the last two are sufficient), followed by another set, commencing with 1, will generally be quite enough; corresponding entries being made in the first column of a ruled diary. These figures may be written in ink on the under side of the card of a mounted specimen, or on a circular disc of paper, pierced by the pin of one too large to be carded.

VI.

HYMENOPTERA.

BY JOHN B. BRIDGMAN.

HAVING been asked to give some instructions as to the method of setting and preserving the aculeate *Hymenoptera*, it is with great pleasure I comply, and I hope it may be the means of inducing others to collect these insects. To begin at the beginning, it is almost needless to state that the females of all of them (a few of the ants excepted) are furnished with stings, but with very little care one need never be stung. As Mrs. Glass says, "First catch your hare": so first I shall give a few instructions where to look for and how to catch these insects. All the apparatus necessary is a gauze ring-net, a cyanide bottle, and a pocket full of small card pill-boxes; the cyanide bottle is best made by wrapping a small piece of cyanide of potassium in two or three thicknesses of blotting-paper, tying it round with cotton to prevent it shaking out, then fixing it to the bottom of a wide-mouthed flat bottle with sealing-wax, which is made to adhere firmly to the glass by heating the glass carefully over a lamp, and then corking it up. The pill-boxes ought to have the tops and bottoms fastened in with liquid glue (a preparation of shell-lac). These are all that are required to catch and bring home the game; which is to be looked for at the flowers of trees, bushes, and plants —one season's experience will teach the best, as some species frequent one, some another, and some almost all. The flowers I have found the greatest favourites are sallows, willows, sycamore, holly, blackthorn, bramble, hawkweeds, ragwort, thistles, and umbelliferæ. Some bore in putrescent wood, and must be looked for on or in the neighbourhood of old posts and palings; some are to be found flying about dry banks, hard-trodden pathways, on heaths, while old sand pits are favourite places; but they should be sought for in any warm, rough, weedy spot; and some may be obtained by digging them out of their burrows with a trowel. My plan of proceeding, after having got one in the net, is to catch hold of the net so that the insect is inclosed in a sort of sack, I then uncork the cyanide, and introduce that into the sack, holding the net firmly round the neck of the bottle, so that there is no other escape for the insect from the net but into the bottle, then gradually work the insect into the bottle and close the mouth with several folds of the net, watch my opportunity and insert the cork:

when the insect is stupefied, which happens in a few
seconds if the bottle is slightly warm, I turn it into
the pill-box. A word of caution : it is necessary to
be methodical in carrying the boxes. I always keep
the empty ones in my right-hand pocket, and the
filled ones in the left-hand one, as, if they are
carried sometimes one way, sometimes another,
sooner or later a previously filled one will be opened
to put an insect in, which will result in the former
tenant speedily making room for the new comer;
and my experience has been, if you do lose anything
it is generally your best capture.

Having got home with the left-hand pocket more
or less filled, turn the boxes out, preparatory to
killing the contents, which must be done with burnt
sulphur. My mode of proceeding is as follows:—
I stupefy the contents of each box with chloroform,
in a manner I will describe farther on. Having
stupefied them, I empty them all into a short, wide-
mouthed, round bottle, having a piece of glass tube
put through the cork; the mouth of the tube is
plugged with cotton wool, not too tight, to act as a
strainer. I then put this in a Nabob pickle-bottle
(any other bottle will do as well), through the
stopper of which I have drilled a hole about a six-
teenth of an inch in diameter, in which is fixed a
copper wire, having a shallow tin cup at the end.

H

In this tin cup is placed the sulphur. The tin cup is then held over the flame of a lamp, gas, or candle, till the sulphur is burning, then put it into the bottle and press it down. When all the oxygen is consumed the sulphur goes out. Leave them for about three hours, take them out, and put them into a damp box for twelve or more hours: they will then be in a splendid condition for setting. To stupefy the insects I tip the lids on one side, put them into the sulphur bottle, pour a drop or two into the tin cup, and put it into the bottle. Be careful not to chloroform them too much, as if killed so they become so rigid that it is with difficulty they can be set.

Having killed them, there only remains to pin and set them. There are various sizes of pins used; most collectors have fancies of their own on this subject; I shall therefore only say what is my practice. The pins I use are D. F. Tayler & Co.'s, New Hall Works, Birmingham; No. 15 for bumble-bees only; the other sizes I find most useful are 15, 10, and 18. Some pin the insects straight, and some with the pin inclining forward. Having pinned them, the next thing is to set them. There are two ways of doing this; one is, cut an oblong square of out cardboard, and put a pin through one end; after the legs are stretched out, this is put into the

cork, one on each side, till the upper surface of cork
is just below the level of the wings, which are then
laid out on the card, and held there by a brace the
same shape as the table (see Fig. 28). If the insect has been pro-
perly killed, the legs
and antennæ will
keep set out without
the aid of pins; if
not, this is done
with bent or straight
pins, as may be ne-
cessary. The other
way is a " hymn
of my own com-
posing."

First take one of
the strips of cork as
sold at the shops,
paper it on both
sides with thin soft
paper; then take a
piece of wood a little larger than the cork, about
half an inch thick; on this I glue strips of cardboard, or thin wood, according to the size of the
insect, side by side, and as far apart as necessary
(see Fig. 29). These being dry, I glue the sheet of

Fig. 28.

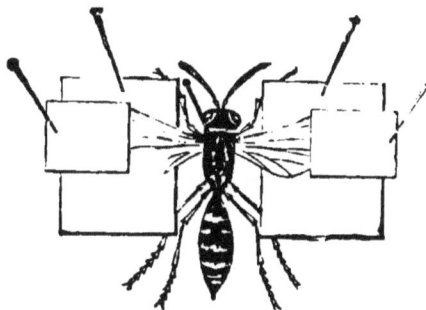

Insect set with Table Braces.

Fig. 29.

Wood, with the Strips glued on.

H 2

cork on to the top of the strips, which leaves it looking like a succession of bridges. When this is dry the cork must be cut through between the pieces first fastened on the wood. These pieces are then taken out and glued to the wood (see Fig. 30);

Fig. 30.

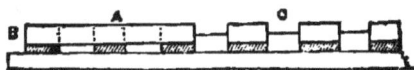

Ditto, side view. A, the same with the cork glued on; B, cork; C, the same with the cork cut through at the dotted lines in A, and fastened down.

this leaves many setting boards, something similar to the single rounded ones used by Lepidopterists; but these are flat—they want to be just deep enough for the insect and wide enough to allow the legs to be stretched out. A little practice will soon determine the size. The wing I hold down with small triangular braces. Each board will hold about seventy or eighty insects; beneath I put the date they were set, and leave them on the board about a month to dry, as if taken off too soon the wings spring. Always put a label to each specimen, either with the date or a number corresponding to one in a book, in which enter the date and locality.

One more observation and I have done. Sometimes one comes across an insect whose rigid wings seem to defy all attempts to set; in such cases just press firmly at the back part of the thorax, between that and the abdomen, towards the pin, and the wings will sometimes fly open of their own accord, or will allow of their being easily set in the required direction, which should always be set well forward.

VII.

LAND AND FRESHWATER SHELLS, ETC.

BY PROFESSOR RALPH TATE, F.G.S., ETC.

A YOUNG friend, desirous of entering upon one of the most accessible natural history pursuits, that of the study of Land and Freshwater Molluscs, begged of me some practical hints on the collection and preservation of these objects of our woodlands, way-sides, and watercourses. Believing that this kind of work offers a good stepping-stone to the study of nature in its more extended forms and complicated relations, I was most anxious to help my tyro naturalist, and that beyond his utmost expectations, as I made a few initiatory trips with him in a search for the coveted treasures.

Our equipment was simple and inexpensive, consisting of a block-tin saucepan finely perforated at the bottom, about six inches across, and having a hollow handle of a size to receive firmly the end of a common walking-stick—such a *dredge* or a *sifter* will cost ninepence or a shilling at a tinman's; secondly, of a pocket lens; and lastly, of a variety of boxes, and a bag to contain specimens of different sizes. Thus provided, our first excursion had for its object

an examination of certain neighbouring ponds and
streams. My pupil, guessing the use of the perforated saucepan, makes his way to the nearest pond,
fixes the improvised handle, dashes in the *sifter* with
impatient ardour, and having brought up a quantity
of mud from the bottom, looked upon the oozy mass
with despair. Patience, my lad! Remember that
the pleasure of success in science is the higher
the greater the labour expended in obtaining the
objects of our search. Expect failure now and again,
but do not be disheartened. *Ohne Hast ohne Rast*,
should be the motto of every naturalist. Now, shake
the tin in the water, keeping its rim just out of the
water, dipping it down now and then. That is well;
thus you see that you have cleared off the mud, and
what you want is probably left behind along with
the rubbish. What, nothing! Come, try again; but
this time scrape the sifter along the surface of the
mud, and I am confident that you will find something to reward you, and with much less trouble and
display of temper. In this way, after repeated trials,
a number of shells were secured and transferred to
the boxes. Then, after the first gush of excitement is
over, we retire to an adjoining bank to con over the
spoils, and I to make mention of the various habits
of freshwater snails, and consequently of the different
modes of search. My young friend's enthusiasm is

aroused by the mention that a few large mussel-like
shells are inhabitants of our fresh waters, and great
is his haste to be up and again doing, in the hope of
adding some of them to his stock. But in vain were
his many attempts to find them in the pond which
had already yielded us such a variety. "Do they
live here?" is at last the anxious question. "No;
but let us away to yon sluggish brook, for it is in
such that we may expect to meet with them." "Now
I see them. Are not those their ends just peeping
above the mud?" And full of eagerness he dashes
in the dredge, but with little result, excepting that of
a dead shell or two. "Oh! how can I get them?
Shall I take off my shoes and socks and wade for
them?" "Well, you might secure them that way,
and sometimes it is the only way, but on this occasion
I do not think it necessary. Come, we will move a
little higher up, where the stream is clear, and the
shellfish undisturbed. Observe the gaping ends of
the shell, and thus I push the end of the stout rod
between the partially-open valves; now they close
upon the stick, and so we bring our prize holding on
to the stick to the bank."

"You will recollect," addressing my companion,
"that in the muddy pond we have just left, we chiefly
got small bivalves and only a few snail shells. I
have already told you that water shells differ much

in their habits, and that consequently our 'search
for any particular species, or set of species, can only
be successfully carried on when that knowledge is
our guide. Those little bivalves, and a few of the
snails that we have gathered, habitually live at the
bottom, and will of course be brought up in the
dredge when that implement is dragged over it;
but there are many shells which live at or near the
surface, and which feed on the submerged and floating plants. Therefore we must seek out a weedy
pool if we would increase the variety of our collection." Such a spot is reached; and the dredge is
brought into requisition, anon to snatch up a floating snail, or again to sweep over and through the
plants, varying our occupation by dragging to the
margin the tangled masses of weeds; by all of
which means a considerable number of the class of
air-breathing water snails was obtained—admonish-
ing my young friend that this last plan does very
well when the plants grow in dense masses, because
when thus interlaced they form a natural net to
catch those snails which on the slightest disturbance
lose their hold upon the weeds, and which would
otherwise fall to the bottom.

Yet another plan remains to be pursued, one by
which the few small shells hiding among the roots
of the plants may be secured. Obviously the dredge

misses such; but by pulling up the plants by their roots, and well shaking them in the half-sunken sifter, we yet after all obtain them.

From causes which need not be explained here, the shells living in some ponds are all much eroded, or coated with a ferruginous deposit; it will be desirable therefore to find out the localities where specimens are in the best condition, so that you may have typical specimens for comparison before an extensive collection is made.

Our experience is, that though a considerable number of species may be obtained from a ditch or pond, yet a few are found as the sole molluscan tenants of particular sheets of water; that lakes exhibit a dearth of life, and that the greatest variety is often to be met with in canals; but should a search be carried on in them, avoid the towpath side, for reasons that a little thought will readily suggest.

Living near the sea, and within a short distance of wooded hill-sides, we had within a limited area such a variety of physical features that we were led to infer the existence of a rich molluscan fauna for the neighbourhood. Our second excursion was devoted to a search for snails along the sea margin and shores of the estuary. Proceeding along the low sand-dunes—at first sight a most uninteresting spot—*Helix caperata, H. virgata, Bulimus 'acutus,* and a few other snails, were found clustering upon the low stunted vegetation in such numbers, that handfuls might have been gathered within an area of a few square feet. Leaving the seashore, our way led us over the foreshore of the mouth of the river, crushing under our feet at every step shells of *Cardium edule, Scrobicularia piperata,* and a few other bivalves which find a congenial habitat in such situations. Gaining the muddy margins of the higher part of the estuary, *Conovulus* was looked for, and found under the stones along the high-water mark. Higher up the river the rejectamentum on its banks was carefully turned over, and we were successful in securing a number of land shells. The animals, of course, do not live in such places; but their empty shells, which alone were found, had been brought down from the land surface by the agency of the streams and tributaries of the river. Nevertheless such an *omnium gatherum* should demand attention, as its contents give an insight into the character of the land and freshwater forms within the area of drainage of the river.

The number of estuarine species which have a place in our works devoted to British land and freshwater snails is very few, and the majority, moreover, are confined to the margins of the tidal rivers in the south of England. Thus *Assiminea Grayana, Hydrobia ventrosa,* and *H. similis,* live on the mud banks beneath the shade of sedges and rushes, skirting the Thames below Greenwich. To gather these small shells singly is a tedious operation; but if a thin piece of flat wood, or other substitute as the ingenuity of the collector suggests, be used to scrape lightly over the surface of mud, transferring the mass to the *dredger,* and washing in water, a number of specimens sufficient to stock every private cabinet in the country may be obtained in a short space of time.

For the third initiatory excursion our steps were directed inland, and as we proceeded the hedgerows, mossy banks, and margins of watercourses were diligently searched, finding a *Helix* here, a *Pupa* or a *Succinea* there. Gaining the woods, we turn over the damp leaves, grub under the clumps of ferns and wood-rushes for small Helices, Pupæ, and the like; scan the trunks of the trees for the climbing *Clausiliæ, Bulimi,* and *Helices,* not unmindful that each little dirt-like mass is probably a *Bulimus obscurus,* which, by covering its shell with mud, thus exhibits a protective faculty, and often escapes detection. Raise the rotting bark for *Balia;* lift the stones at our feet, or roll away a log for *Helicella,* and other small shells which usually live in' such situations.

From all this we learn that each species affects certain stations, and therefore, with the knowledge of the circumstances in which they are found, we may set out with some definite idea as to what we are likely to meet with; and, in consequence, when to collect and where to collect are regulated by the unvarying habits of the objects of our search.

Now, a large portion of the life of most land snails is passed in a state of sleep. Those living in open situations are inactive during the heat of a summer's day, and when there is continued drought; but on the first shower, or after the fall of dew at night, they recover and move about in search of food. Cold acts much in the same way as heat, and with the fall of the leaf they retire to winter quarters in crannies of rocks, crevices of walls, under heaps of decaying vegetation, &c., or bury themselves in the soil, there to hybernate till the genial showers of spring awaken them.

The best time of the year for collecting is in the autumn, when the shells are full-grown. Those collected in spring have lost much of their original beauty by exposure to the rains and cold of the winter months.

As regards the particular time of day to collect with advantage, it has already been implied that a search in an open country should be prosecuted after a shower of rain, or during early morn. In damp woods, where throughout the day the air is sufficiently moist to maintain the animals in full activity, no such considerations determine the best time for collecting. In such places, light is usually the desideratum, and consequently I have found that a search conducted at midday in a clear sky has been amply rewarded.

Land snails exhibit a partiality for calcareous soils, not only by those living on downs and hillsides, but also by the woodland species.

Having spent the forenoons of three days in gathering slugs and snails as before detailed, one evening was devoted to the preparation of the specimens for the cabinet.

The first step was to remove the animals, and, as all know, it is neither an easy nor a cleanly task to separate the living snail and its house; but kill your snail, and the muscular connection with the shell being severed, its whole body is readily taken out by means of a pin—why, it is just like picking periwinkles; and if the proclivities of our childhood's days are not entirely obliterated, cleaning out larger snails from their shells will be a task re-

quiring no teaching.   But, with regard to the smaller
kind, it is another matter, and it will be my duty to
show you how to set about the work.

Now pick out those shells, the apertures of which
are wide enough, as it seems to you, to permit the
removal of the dead body of the snail by a pin. You
may also place with them the larger bivalves. All
these we will boil to kill the animals ; then strain
off the water, and wash with cold water. By this
means the bodies contract, and being firmer are not
so liable to be broken in the process of removal.
Shake the water out of the empty shell, and place
them before the fire to dry; do not rub them, but
particles of dirt may be gently flicked off by the aid
of a camel-hair brush. Thus we treat the larger
snails. Now for the mussels. Doubtless most of
the dead bodies will have fallen out between the
open valves while in the water ; should any remain,
a slight shaking of the shell held by the hand in the
water will remove the contained body. Taken from
the water, the valves gape widely ; dry the inside
and outside with a cloth, and having tape or cotton
at hand, close the valves by the pressure of the
thumb and fingers of the one hand, and with the
end of the thread between your teeth, wind the
thread two or three times around the shell with the
other ; now tie the thread as tight as you can. " Yes,

I have done so, but still the valves are not closed." True, this is because of the elasticity of thread. If, however, you will take the precaution to wet the thread before tying, you will find that the tie is more secure, and that there is less difficulty in making the second knot.

With patience and a little skill, bivalves as small as *Cyclas cornea* may be treated in this way. But the smaller *Pisidiums*, and some of the minute snails, as *Carychium minimum*, may be prepared for the cabinet by gently drying them in sand ; too great a heat causes a transfusion of the carbonaceous matter of the animal into the substance of the shell, which is thereby discoloured.

There still remain for treatment such shells as *Clausilia, Bulimus, Helicella*, some Helices, &c., the animals of which retreat, on the least irritation, beyònd the reach of a pin, and whose shells, indeed, will hardly bear the rough handling almost necessary when a pin is used. Their bodies might be dried within the shells, but if it be possible to remove some portion only of the animal, an attempt should be made to do so.

Land snails, when placed in water, do their best to effect an escape from a medium so fatal to them ; their efforts are usually exhibited by stretching out their bodies to the utmost, swaying them to and fro as if in search of a foothold. Taking advantage of this propensity, the snails should be immersed in tepid water, because the majority, after a day or two's confinement in the collecting boxes, will be in a dormant condition, and warm water has a greater resuscitating effect than cold. When all the snails are struggling to find a way out of their unpleasant situation, gradually add hot water so as to kill or paralyze them while in an extended state. They may now be thrown into boiling water, the better to relax the muscular attachments, and the bodies, or so much as will come away, dragged out by forceps, or a pin passed through the foot. The shells may now be dried in sand, as before mentioned.

In cleaning the shells of some species, great care is needed, so as not to remove the hairs or bristles which clothe the surface of the epidermis.

The shells of such snails as *Paludina, Cyclostoma,* &c., &c., would be imperfectly illustrated without the opercula or lids which close the apertures of their shells. Each one should be detached from the foot of the snail, the interior of the shell plugged with cotton wool, and the specimen gummed down in its natural position.

The preservation of slugs requires separate treatment, and I can give but little additional information to that published in my 'British Land and

Freshwater Molluscs,' an extract from which is subjoined :—

" As regards the internal shell, it may be obtained by making a conical incision in the shield, taking care not to cut down upon the calcareous plate, which can then be removed without difficulty. The animals can only be conserved by keeping them in some preservative fluid; but the great object to keep in view is to have the slug naturally extended. Most fluids contract the slugs when they are immersed in them. The slugs should be killed whilst crawling, by plunging them into a solution of corrosive sublimate, or into benzine. Models in wax or dough are sometimes substituted for the animals. A writer in the 'Naturalist' gives a process for the preservation of slugs, which he states to answer admirably, and to be very superior to spirit, glycerine, creosote, and other solutions :—' Make a cold saturated solution of *corrosive sublimate;* put it into a deep wide-mouthed bottle ; then take the slug you wish to preserve, and let it crawl upon a long slip of card. When the tentacles are fully extended, plunge it suddenly into the solution; in a few minutes it will die, with the tentacles fully extended in the most life-like manner; so much so indeed, if taken out of the fluid it would be difficult to say whether it be alive or dead. The slugs thus prepared should not be mounted in spirit, as it is apt to contract and discolour them. A mixture of one and a half parts of water and one part of glycerine, I find to be the best mounting fluid. It preserves the colour beautifully, and its antiseptic qualities are unexceptionable. A good-sized test-tube answers better than a bottle for putting them up, as it admits of closer examination of the animal. The only drawback to this process is, that unless the solution is of sufficient strength, and unless the tentacles are extruded when the animal is immersed, it generally, but not invariably, fails. Some slugs appear to be more susceptible to the action of the fluid than others; and it generally answers better with full-grown than with young specimens. But if successful, the specimens are as satisfactory as could be desired; and even if unsuccessful, they are a great deal better than those preserved in spirit; for, although the tentacles may not be completely extruded, they are more or less so.' "

The Testacellæ I have treated in the following manner: by partially drying them in sand, and removal of the soft parts through a cut in the length of the foot, filling up with cotton wool and a renewal of the drying.

Our land and freshwater snails have other structures besides their shells which should claim our

I 2

attention. These, which include their jaws, tongues, and some other minute parts, are not so inaccessible as one is at first too apt to consider, and are deservedly in favour as microscopic objects requiring a low power. I shall assume that the collector has preserved the bodies or the heads of the snails in spirit, which he has removed from their shells in the process of preparing them for his cabinet. He will take care to keep separate the animals of each species.

A last word upon the mode of displaying the shells in the cabinet. Here one has considerable choice, as they may be kept in open card trays, or in glass-topped boxes, or gummed on cards, papered boards, or glass tablets. Loose specimens admit of ready examination, whilst the method of mounting permits an arrangement of individuals according to size and locality, and is much to be preferred.

VIII.

FLOWERING PLANTS AND FERNS.

BY JAMES BRITTEN, F.L.S.

PART I.

THE kindred subjects of the collecting of plants and their arrangement in the herbarium have been treated of over and over again, and it might almost seem as though nothing further need be said upon the matter. But in spite of all that has been written, it cannot be said that anything like uniform excellence has been attained, either in the collecting or drying of specimens: on the contrary, much carelessness is still exhibited in both particulars, and the following remarks on the subject may therefore be useful to some, at any rate, among the readers of 'Science Gossip.' It has been found impossible to treat both points adequately in one paper, so, on the present occasion, we shall devote ourselves to collecting, leaving the arrangement and matters connected therewith for another occasion.

The great aim to be kept in view in collecting is to obtain as perfect and comprehensive a specimen as possible; that is, one showing every part of the plant—root, leaves, flowers, and fruit. It is not

Fig. 31.

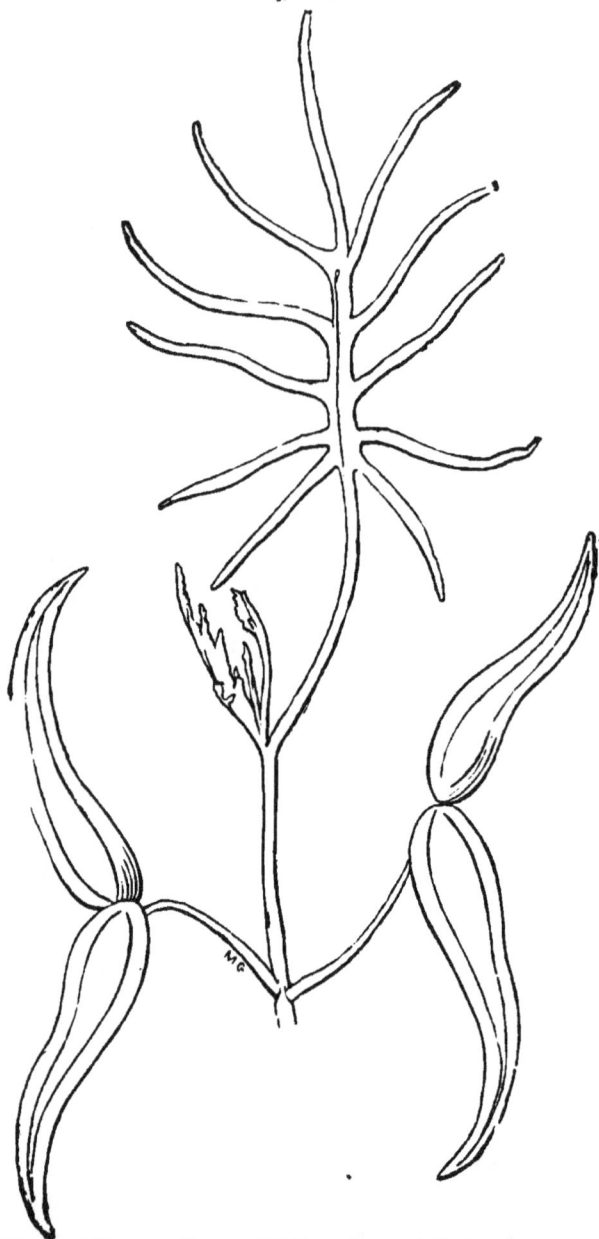

Young Plant of *Ipomœa Quamoclit* (from Decandolle's ' Organographie ').

always practicable to show all these upon one specimen, and in such cases such a number must be
selected as will carry out this plan. The wretched
scraps with which some collectors content themselves
are not only useless to their owners, but annoyances
to everyone who has to do with them, or who is
requested to pronounce an opinion upon them.
Anyone who has had anything to do with naming
plants for 'Science Gossip,' or any other journal,
which in this manner supplies information to its
subscribers, will be able to testify to the large
number of persons who do not scruple to send for
determination single leaves, or a terminal shoot of a
flowering plant, or a pinnule of a fern without fruit;
a proceeding which is unfair to those to whom they
are submitted, inasmuch as they either have to risk
their reputation for accuracy, or to appear un-
courteous by refusing to have anything to do with
such specimens.

To begin at the beginning, How rarely do we find
the embryo of any species represented in a collection
of dried plants? It ought to be there, not only as
essential to the complete presentment of the history
of the species, but as in certain cases indicating
relationships which are not apparent when the plant
is more advanced. Those who have not observed
them would be surprised to find how much variety
of form exists in the cotyledons alone, from the

fleshy cotyledons of many of the Leguminosæ, the horse-chestnut, &c., to the foliaceous ones, or seed-leaves, of other plants. Among the latter may be noted and compared the lobed or palmate cotyledons of the lime (Fig. 32); the glossy dark green, some-

Fig. 32.

Lime (*Tilia Europæa*).

what kidney-shaped ones of the beech (Fig. 33); and the pinnatifid ones of the common garden cress (*Lepidium sativum*); the obcordate ones of the mustard or radish; the long, narrow, thin ones of the

sycamore (Fig. 34); the sinuous or corrugated and
bilobed ones of the walnut, and many more which will

Fig. 33.

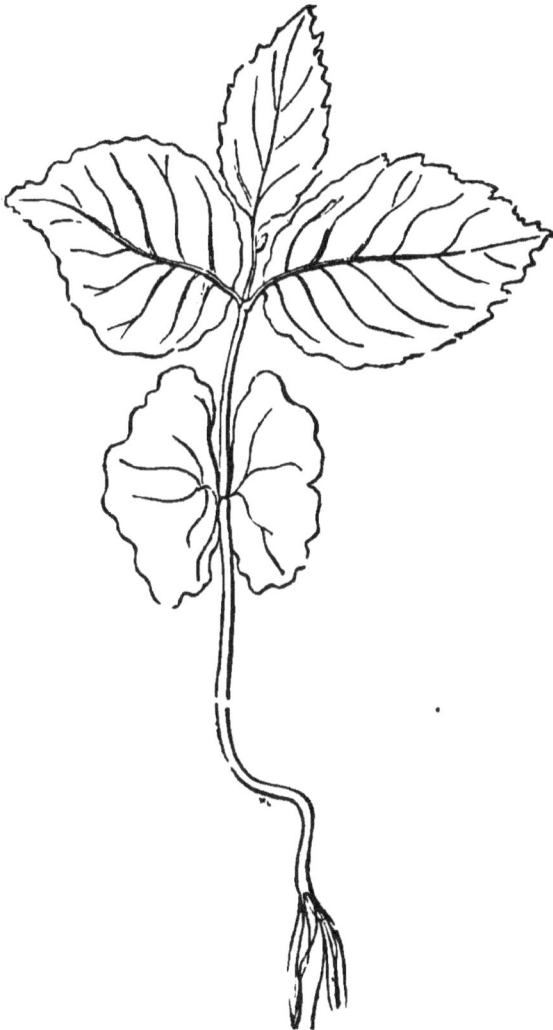

Beech (*Fagus sylvatica*).

occur to the observant reader, or which may be collected by anyone who will take the trouble to watch
the germination of plants. And by making such
collections, unexpected discoveries may arise, which

Fig. 34.

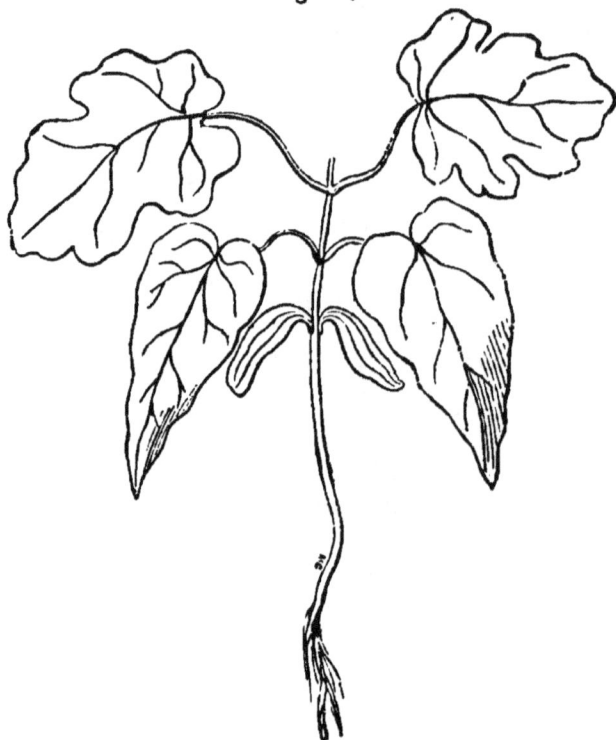

Sycamore (*Acer pseudo-platanus*), showing cotyledons and first and
second pair of leaves.

will yet further confirm what has been said about
the variety in form and structure even in these
beginnings of growth. Plants which are, on ac-

count of their general affinities, reckoned among the dicotyledons, may be found on investigation to have but one cotyledon, as Dr. Dickson observed to be the case with two of our butterworts, *Pinguicula vulgaris* and *P. grandiflora*, the third species, *P. lusitanica*, being dicotyledonous; or even to be acotyledonous, as is the case with the dodder (*Cuscuta*). In the latter-named genus, it is of importance to collect young specimens, as showing that the plant, although parasitic as soon as it comes in contact with a suitable foster-plant, is of independent origin. A search among young plants will no doubt lead to the discovery of some abnormalities, such as tricotyledonous embryos and other irregularities. Of some plants, such as the furze, the true leaves can only be found at an early stage of growth; in others, much variation may be noted in many points between the first leaves and the more perfect ones which succeed them; some, as the holly, at once developing leaves similar to those which are produced throughout the life of the plant, and others going through many modifications before the ultimate shape is attained, as in the ash, elder, ivy, maple, &c.

The roots or rhizomes also require to be much more fully represented and carefully collected than is usually the case. In every instance where the size of the plant does not prevent, the subterranean

and subaqueous parts should be carefully procured
and preserved.

Dr. Trimen has lately directed attention to the
corm-like tubers of the water plantain (*Alisma*),*
closely resembling those of the arrowhead (*Sagittaria*), which have been described and figured by
Nolte, but "do not seem to have been observed, or
at least properly understood, in this country. They
are buds remaining dormant through the winter,
and containing a store of nutriment, to be employed
in the development of the new plant from the tuber
in the next year." Similar bulbs are developed by
the frogbit (*Hydrocharis*). In determining many
grasses and rushes, it is of importance to ascertain
whether the rhizome is creeping or cæspitose, and it
is therefore essential to collect good specimens. In
the case of such plants as the coral-wort (*Dentaria
bulbifera*) and toothwort (*Lathræa squamaria*), the
root-stocks are eminently characteristic. Of such
parasites as the broomrapes (*Orobanche*), some care
is requisite in obtaining specimens in which the
connection between the parasite and its foster-plant
may be preserved and shown. The absence or presence of tubers should also be noted, and if present,
they must be represented.

Passing on to the leaves, we may note the im-

* 'Journal of Botany,' 1871, p. 306.

portance of obtaining in every case the root-leaves of each species. These are often very different in form from the stem-leaves, as in such species as the harebell (*Campanula rotundifolia*), *Pimpinella saxi-fraga*, the earth-nut (*Bunium flexuosum*), and many more; in some instances, as in the Jersey bugloss (*Echium plantagineum*), they at once characterize the species. Still more important are these lower leaves in the case of water-plants: in the arrowhead (*Sagittaria*), for example, they are narrow, and re-semble those of the bur-reed (*Sparganium*); and in the water plantain (*Alisma plantago*), the submerged leaves are equally different from those which rise out of the water. This difference is still more noticeable in the case of the aquatic *Ranunculi*, where a know-ledge of the submerged leaves is essential to the discrimination of the various forms or species.

Where practicable, the whole plant should be collected for the herbarium; but when, from its size, this cannot be accomplished, leaves from the root, the centre of the main stem, and the lateral branches, should be taken. As to the stem itself, that must be represented: in the *Rubi*, indeed, it is essential. " To judge properly of a bramble from a preserved specimen," says Professor Babington, " we require a piece of the middle of the stem with more than one leaf; the base and tip of the stem are also desirable, likewise a piece of the old stem with the flowering shoot attached to it; the panicle with flowers, and the fruit. We likewise want to know the direction of the stem throughout, of the leaflets, and of the calyx; also the shape of the petals and the colour of the styles: a note of these should be made when the specimen is gathered."

Passing on to the flowers, we shall find it necessary to represent them in almost every stage, from the bud to the perfecting of the fruit. It is of course in most cases possible to select an example in such a state as to show upon the same plant buds, flowers, and fruits; but where this is not the case, each of these particulars must be supplemented by additional specimens. The turn which botanical investigation has recently taken towards the study of the phenomena connected with fertilization has given the collector another subject to which his attention may be profitably directed. It has been observed that in some plants the stamens are developed before the pistils; in others, the pistils are matured before the stamens; while in yet a third set, stamens and pistils are simultaneously perfected. These three groups of plants are termed respectively protandrous, protogynous, and cynacmic, and a very little observation will show that examples of each are sufficiently common.

Then in diœcious and monœcious plants, both
male and female flowers must be collected, and in
some cases, as in the willows, four specimens are
necessary to the complete presentment of the species,
showing respectively the male and female catkins,
the leaves, and the fruit. Some plants produce two
distinct forms of blossom, as is noticeable in the
violets and the woodsorrel, one being conspicuous
and usually barren, the other insignificant and often
apetalous, but producing perfect fruit. The pollen
will afford occupation to the microscopist: the researches of Mr. Gulliver and Mr. Charles Bailey
have demonstrated that important distinguishing
characters are in some instances furnished by it.
While on this point it may be suggested that it is
convenient in many cases to collect several specimens
of the flowers alone, which, when dried, should be
placed in a small envelope or capsule, and attached
to the sheet on which the plant is represented. In
the event of any examination which may be requisite
after the plant is dried, these detached blossoms will
be found very useful, and will prevent the necessity
of damaging the specimen. In the case of such
plants as shed their corollas very readily, as the
speedwells, it is as well to put them in press as
soon as collected ; and the colour of many may be
retained by the same means.

The fruits and seeds of plants are too generally
neglected by amateur collectors, but are essential to
the completeness of a specimen. It may be found
practically convenient to keep these in a separate
place, and detached from the plant; and in many
cases of dried fruits it is advisable to sort them into
their places without previous pressing. By this
means the modes of dehiscence will readily be seen:
pulpy and succulent fruits should be preserved in
spirit. In such plants as the species of sea sandwort
(*Lepigonum*), and some Chenopodia, important specific
characters are drawn from the seed; as they are
from the pods of *Melilotus* and the fruits of *Agrimonia*.
In collecting ferns, well-fruited fronds must be
selected, as it is impossible to determine specimens
without fructification. Grasses should be selected
when in flower and fruit, but must not be allowed to
attain too great an age before they are collected.

We have been speaking so far of the things to be
collected, and space will not allow us to dilate at
any length upon the apparatus necessary to that
end. Nor indeed is this necessary; a good-sized
vasculum, with one or two smaller boxes for the
pocket, in which the more delicate plants may be
preserved; a strong pocket-knife or small trowel,
for procuring roots, and a hooked stick wherewith to
fish out water-plants, or pull down branches, are the

principal things required.   To anyone residing for
any length of time, or even only for a few days, in
a district, a "London Catalogue" is an important
acquisition, in which should be marked off all the
species met with; by this means the flora of the
neighbourhood is ascertained at a very slight expenditure of time and trouble.   It is not advisable
to collect too many plants at once, or to crowd the
vasculum, unless under exceptional circumstances;
nor should the desire to possess rare plants tend, as
is too often the case, to the neglect and exclusion of
commoner ones.

A careful and observant collector will frequently
meet with forms which deviate more or less from
the accepted type of a species.   When these appear
to offer any marked characters, they should be noted;
and in all cases it is well to preserve any forms
which, from external circumstances, have a different
appearance from the normal state.   The differences
produced by soil and situation alone are very considerable; and though the essential characters are
usually to be discerned, the interest and value of a
herbarium are very much increased by a selection of
examples showing the range of a species.   *Campanula
glomerata* offers a good example of this.   In damp
meadows it is from one to two feet high, with a
large spreading terminal head of blossoms, while on

chalk downs it does not attain more than as many inches, with only one or two flowers; in this state it was described by Withering as a gentian, under the name of *Gentiana collina;* and the same author gives as *Campanula uniflora* a one-flowered mountain state of the harebell (*C. rotundifolia*).

The collector will also do well to keep a look-out for deviations in structure, which are often of great interest. In short, nothing should be neglected which can tend to the perfect presentment of a species in the herbarium: its utility is commensurate with its completeness. The mere collector may be satisfied with scraps of a rare plant and the absence of commoner species; but the real worker will pride himself rather upon the possession of instructive examples, which may be of assistance to himself, as well as to all those who may consult them.

IX.

FLOWERING PLANTS AND FERNS.

BY JAMES BRITTEN, F.L.S.

PART II.

WE will assume that our collecting for the year has come to a close; that the long evenings are beginning, and that our dried plants have been brought together from their temporary resting-places to be revised and selected from, so that they may be intercalated in their places in the herbarium, if we already possess one, or, if we are as yet quite novices, that they may form a nucleus around which the whole British flora shall be gathered in due course. First of all, we must make all necessary preparation for—

Mounting, the first essential to which is paper. Much of the neatness of a herbarium depends upon its uniformity, so that it is desirable to lay down a definite plan at the beginning and to act up to it consistently. Amateurs often spoil specimens which they have collected and preserved with considerable care by transferring them from one sheet to another; from books—but it is only *very* amateur botanists

K 2

who keep their plants in this way!—to loose sheets,
from small paper to large, and so on; each change
being attended with some slight damage to the
specimen so treated.   It is, I believe, the common
practice on the Continent to keep the specimens
loose in folded sheets of paper; but this plan is not
followed in England, and although advantageous, as
permitting the fullest examination of the plant, it is
attended with much risk to the specimens in the
way of breakage; so that we may consider it settled
that we are going to fasten our plant down upon a
sheet of paper.   This must be rather stout, and large
enough to admit the full representation of the
species.   The sheets ·used at the Kew Herbarium
are 16½ inches long by 10⅓ inches wide; those em-·
ployed at the British Museum are 17½ inches by
11¼ inches; but the former will be found amply
sufficient for our purpose.   The next consideration
is the means by which the specimens are to be
secured, which are more various than might at first
be supposed.   Some persons sew them to the paper;
others p'ace straps over them, which are secured
with small pins; but the choice practically lies
between fixing.the whole specimen to the paper with
gum, paste, or glue, or securing it with straps of
gummed paper.   The former plan, which is that
adopted at our great public herbaria, is certainly

better for specimens which are likely to be much consulted; but the latter is in some respects more satisfactory, if somewhat tedious, as it admits the removal of the plant to another sheet if necessary, and delicate portions, such as thin petals or leaves, are not injured as they are when gummed down. At the British Museum and Kew a mixture of gum tragacanth and gum-arabic (the former dissolved in the latter), in about equal parts, is used for this purpose; but very coriaceous specimens are secured with glue at the last-named establishment, while in the former the stems and ends of branches are usually also secured with straps. When the specimen is entirely gummed down, it is a good plan to keep a few extra flowers or fruits in a small capsule attached to the sheet: these will be useful if it is required to dissect such portions, and the specimen need not be injured for such purpose.

Poisoning.—Some persons are in the habit of employing a solution of corrosive sublimate for the purpose of washing over their plants when mounted, and so preventing the development of animal life. The solution in use at the Kew Herbarium is composed of one pound of corrosive sublimate, and the same quantity of carbolic acid to four gallons of methylated spirit; this fulfils the purpose for which it is intended very well, but is somewhat disagreeable to use. At the British Museum it is found that the presence of camphor, frequently renewed in each cabinet, is sufficient to prevent the attacks of insects. It will soon be discovered that some plants, such, for example, as the *Umbelliferæ* and *Grossulariaceæ*, are peculiarly liable to such attacks; and these orders must be inspected from time to time, so that any insect ravages may at once be checked. Damp is to be avoided in the situation of the herbarium, as it favours the development not only of insects but of mould, and renders the specimens rotten.

The question of *labelling* is of some importance, especially to those who value neatness and uniformity in the appearance of their herbarium. One or two sets of printed labels for this purpose have been issued, but they cannot be recommended. They give more than is necessary, e. g. the English, or, more correctly, the book-English names, the general habitats, and definite localities of rare species, and allow very insufficient space for filling in the date and place of collecting, the name of the collector, and such remarks as occasionally occur. The plan of writing all necessary information upon the sheet itself is a good one; but those who prefer a uniform series of labels will find that a form like the following is as useful as any which they can adopt, and includes all necessary information. The size here given will be adequate for almost all requirements, and is a "happy medium" between the small tickets upon which we have animadverted, and the enormous ones with which some botanists think it necessary to accompany their specimens. Care should be taken to avoid the possibility of a misplacement of labels; many serious blunders have arisen from the neglect of due precaution in this matter.

Herb. John Smith.

Ranunculus acris, L.
& R. Steveni, Reich.

Loc. Meadows near Barchester.

DATE, June 30, 1874.

COLL. John Smith.

Arrangement.—The plants, being now affixed to their respective sheets and duly labelled, are ready to be placed in covers, and rendered available for ready reference. Each genus will require a separate cover, which may well be of somewhat stouter paper than that on which the plants are mounted; the name of the genus should be written at the left-

hand corner, followed by a reference to the page of
the manual by which the plants are arranged, or
to the number which it bears in the "London Catalogue," if that be employed in their arrangement—a
purpose for which it is very suitable. Should the
species be represented by more than one sheet, it is
convenient to inclose each in a cover of thinner
paper, which may bear the number assigned to the
plant in the right-hand corner; and it is also convenient to write the name of the plant at the bottom
of each sheet, and to number it also in the right-
hand corner. These details may appear trivial, but
they in reality affect in no small degree the readi-
ness with which any species may be referred to.
Should the plants be arranged in accordance with
the "London Catalogue," a copy should be kept with
the herbarium, in which the plants should be ticked
off, so that it may serve as a catalogue of the species
represented.

Cabinets.—It will of course be necessary to provide
some accommodation for our specimens, and for this
purpose we shall find no better model than the
cabinets in use in the Botanical Department of the
British Museum. The accompanying figure (drawn
to scale) is an exact representation of one of these.
The measurements can of course be modified so as
to suit the size of the herbarium sheets. Each shelf

is a separate drawer, which with its contents can be
taken out and replaced at will. Two cabinets such

Fig. 35.

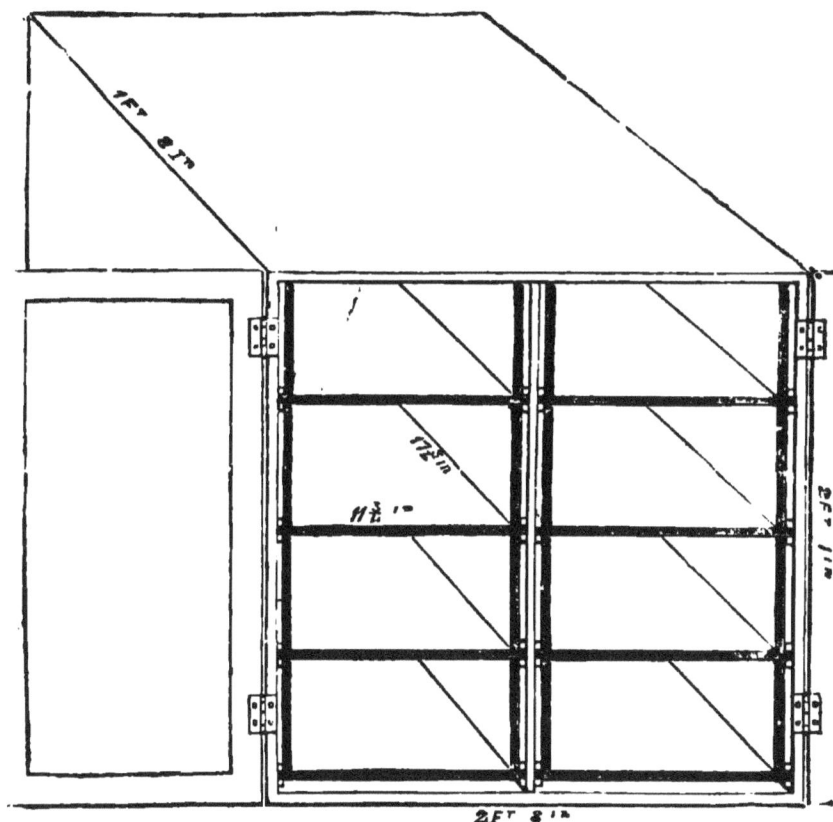

Cabinet for Herbarium Sheets.

as that figured will be found amply sufficient to
contain a very good British herbarium. At Kew
the cabinets employed are somewhat similar, but
their height is greater and the shelves are fixed.

The above are the principal points connected with the arrangement of a herbarium, considered as distinct from the work' of collecting. It is possible that I may have omitted to touch upon certain details which may occur to the amateur; should such be the case, I may add that I shall be happy to supply any additional information, either by letter or by word of mouth; or to show the system adopted at the British Museum to anyone who may call upon me there for further hints upon the subject.

X.

GRASSES, ETC.

By Professor Buckman, F.G.S., etc.

Grasses form such a distinct group of plants, and
their study is so often undertaken for special purposes, that a few remarks upon their collection and
preservation can hardly be considered as out of place
in this little manual.

Delicately as grasses are formed, yet it cannot be
said that their tissues are so liable to injury, or their
colours so evanescent, as those of the flowering plants.
which the botanist ordinarily delights in. Nor indeed are the grasses so succulent as many other
herbs. In this respect they may be said to hold a
place between ferns and those plants which usually
are called flowers.

Again, in the dried state their organs are generally
so well preserved as to present all that a botanist can
wish for, for identification as well as arrangement;
and the student of grasses ever finds his collection
to contain beauties not only in point of rarity, but
as regards delicacy of structure and grace of outline.

Viewing them in this light alone, we have often

been astonished that so many students of plants pay so little attention to them, and this feeling is enhanced when the great value of the grasses is considered.

If then a few simple directions for preserving these plants shall have the effect of winning a convert to these views, we shall be delighted; and to this end we shall make our descriptions as plain as our process has ever been easy and simple, and yet complete.

In collecting grasses, as in other tribes of plants, it will be necessary that our specimens should be chosen with the view to exhibit every feature of interest. With this aim, then, it will be best in the general way to obtain as much of the plant as possible, so that it may be necessary to get them up by the roots. Still, in many species the root is not of much importance: but there are a few which possess *rhizomata*, or underground stems; such as the *Triticum repens, Poa pratensis, P. compressa, Holcus mollis, Agrostis stolonifera*, and others. These should always exhibit these parts; and as such examples are usually agrarian, it is easier to mark down desirable specimens and seek a fork at the neighbouring farm-buildings wherewith to completely get them out, than to carry any substitute in a smaller and less perfect implement.

Having made these remarks, we will suppose that we are now about to sally forth in search of grasses; in which case we make the following preparations.

As we do not file our copy of the 'Times,' we make use of it as collecting-paper as follows :—Each side of the paper is cut in two, or, as a Cockney would say, " in half." Each half is then folded into a double collecting-sheet, and as many of these are taken as are likely to be useful. In each of these papers is put a small slip of writing-paper, on which to note the locality and any other noteworthy fact connected with a specimen when put in the paper. These papers, separately folded, are placed with the open ends inwards in a convenient portfolio, and the collector is ready to take the field.

Of course there will be those who will advocate Bentall's drying-paper, blotting-paper, and so on, and we would not have it supposed that we despise these luxuries; but as we have found the plan advocated always to answer the purpose for grasses, we have felt independent of the more refined collecting-papers.

Now let us suppose that we have gathered fifty specimens, and have returned home. The next thing will be to put them as soon as may be in a position for drying.

Our drying apparatus then consists of half-a-dozen

smoothly planed deal boards, and for our first collection we take two of these, and upon one we lay some few folds of our old 'Times,' then a specimen in their papers (having previously improved their arrangement, when necessary), and then some more folds of paper, and proceed as before, until all the specimens have been placed; then put a board on the top sheet, and upon that a stone, or a 7 or 14 lb. weight, according to the size and quantity of the specimens. If another day's collection of specimens be made before the foregoing are dry, they may be arranged in the same way on the top board, and another board used and the weight replaced. The object of this is to keep partially dried from fresh specimens, the putting together of which is a fertile source of mildew and decay.

In arranging our specimens for the herbarium, we procure sheets of cartridge paper 18 inches long by 11 inches wide, using a folded sheet for each species.

In these papers the specimens are fastened down in the following manner:

Gum over a portion of the cartridge paper (so as to have the same colour) with two consecutive coats of a clean solution of gum-arabic.

This can be cut into slips of any length and breadth, making them as narrow as possible for the sake of neatness, and when the specimen is placed in its paper, a few of these slips may be made to confine it in the desired position. Each example is then to be labelled at the bottom of the sheet, and each label should set forth—*a*, Its botanical name; *b*, its trivial or local name; *c*, the locality whence it was obtained; *d*, the date when gathered; added to which, if presented, the donor's name.*

The sheets so prepared may be arranged in groups or genera, each being folded in convenient paper or cloth wrappers, and the whole arranged in volumes of stiff covered portfolio.

This, then, is all that seems to us necessary in the collection and preservation of grasses; but we would recommend the student, if an artist, to make a typical specimen of each sit for its portrait. In this way we have made drawings of all the species and varieties that have come in our way.

Our drawings are life-size, usually lined in with Indian ink with a fine " lithographic pen." These we partially colour on the spot.

The anatomical details are much enlarged and always fully coloured. To this end our *impedimenta* for a day among the grasses consist of, besides the collecting portfolio, a sketching block, large octavo size, and a small box of soft colours. Armed with

---

* Printed herbarium labels may be got at Messrs. Hardwicke and Bogue's, the publishers.

these we have made many a drawing of a grass under the shade of a tree, or in the parlour of some contiguous inn.

Lastly, we would venture to remark, if, besides the interest which grasses should have for the student of botany, these plants be viewed, as they have ever been by us, as indicators of the nature of soil and the value and capabilities of the land on which they grow, the collector should not fail to make notes connected with the soil, situation, and other practical facts connected with the habitats of GRASSES.

XI.

MOSSES.

BY DR. BRAITHWAITE, F.L.S., ETC.

IN making a collection of the vegetable productions of a country we find considerable differences in the structure of the various groups of plants, and in the tissues of which they are composed ; and hence special manipulation is requisite in dealing with certain orders. Some are of so succulent a nature, or have a framework so easily disintegrated, that they contain within themselves the elements of destruction, and present the greatest difficulty in satisfactory pre-servation, while others are so slightly acted on by external agents, that little trouble is required to prepare specimens of permanent beauty.

The Ferns and Lycopods, being generally appro-priated by the collector of flowering plants, will be treated on with the latter, and following these come the Mosses, to which we will now direct attention, taking the alliance in its broadest sense, as including the three groups of Frondose Mosses, Bog Mosses, and Liver Mosses, or Hepaticæ, all of which are readily

L

collected and preserved, and yield an endless fund
of instructive entertainment to the microscopist.
But it may be asked, Where is the game to be
found? Where are the pleasant hunting-grounds in
which they most do congregate? We answer, everywhere may some species or other be met with; yet,
though many are cosmopolitan, the majority have
their special habitats, and some their special seasons,
both being considerably influenced by the presence
of moisture.

Collecting.—The bryologist has one advantage over
the phænogamous botanist, for it is not impera-
tive that mosses should be laid out and pressed
immediately; and hence less care is required in
collecting them, than is bestowed on flowering
plants; the necessary apparatus is confined to a
pocket-knife, to remove specimens from stones or
trees, a stock of stout waste paper, and a vasculum,
or, better still, a strong bag, in which to carry the
packets. When collecting the plants, it is well to
remove any superfluous earth or stones, or to squeeze
out the water from those found in bogs; and then
each is to be wrapped separately in paper, and the
locality marked outside; or the more minute species
may, for greater safety, be placed in chip boxes.
On reaching home, if we do not prepare the specimens at once, we must not leave the parcels packed

together in their receptacle, or mould will soon attack them and spoil the whole; but we must spread them out on the floor until quite dry, and then reserve them to a convenient opportunity to lay out; as in the dry state they may be kept for years unchanged.

It often happens that our line of study is developed by some fortuitous circumstance. A neglected flowerpot in the corner of the garden attracts attention by its verdant carpet of moss, or, peeping over the wall, we see the crevices between the bricks bristling with capsules of *Tortula muralis*, the red twisted peristome freshly brought to view by the falling away of the lid, and, taking a bit indoors to submit to the microscope, we are so captivated therewith that we then and there determine to become a bryologist. Nor is this all that a journey round the garden will disclose: the neglected paths yield other species not less worthy of examination, and old apple-trees are not unfrequently tenanted by mosses.

Extending our walks to the commons, lanes, and woods, we may find on the ground and banks, in bogs and on the stumps and trunks of trees, a number of species greatly extending our list ; while others again are only met with on the clay soil of stubble-fields, as various species of *Pottia* and *Ephemerum:* appearing in October, their delicate texture

L 2

is developed by the constant moisture of winter, and with it also they vanish, to appear no more until the succeeding season. Travelling yet farther away, we find that each locality we visit yields some novelty: old walls and rocks of sandstone or slate, limestone districts, and, above all, a mountainous country, are rich in species we seek in vain elsewhere. Here peat bogs, and rocks dripping with water, ever supplied by the atmosphere, or the tumbling streams everywhere met with, are the chosen homes of these little plants, and thither must the collector resort, if he would reap his richest harvest. Winter and spring in the lowlands, and a later period in the elevated districts, will be found most productive of fruiting plants.

Preparation of Specimens.—So rapidly does the cellular texture of the mosses transmit fluid, that, when soaked in water, we see them swell up and expand their little leaves, and in a short time look as fresh as when growing; hence a basin of water, a towel, and drying-paper are all we require to prepare our specimens for the herbarium. If the tufts are large, we must separate them into patches sufficiently thin to lie flat, and by repeated washing, get rid of adherent earth, mud, or gravel. This is conveniently accomplished by holding the tuft in the palm of the hand, under a tap, and allowing a stream of water to pass through it; then by pressure in the folding towel we remove superfluous moisture and immediately transfer to paper, arranging the plants as we wish them to lie permanently, and placing with each a ticket bearing the name: a moderate weight is sufficient to dry them, as with great pressure the capsules split, and thus the value of the specimen is decreased. It not unfrequently happens that two or three species grow intermixed: these must be carefully separated at the time of soaking, and any capsules required to show the peristome must also be removed before the plants are submitted to pressure.

Examination of Specimens.—We have very much to learn about a moss before we can become masters of all the characters that pertain to it as a specific individual. We must observe its branching, the mode of attachment of the leaves to the stem, and their direction; the form and structure of a separate leaf, the position of the male flowers, and, lastly, the position and structure of the fruit. For the efficient determination of these we require a microscope (the simple dissecting microscope is amply sufficient), a couple of sharp-edged, triangular needles fixed in handles, and a few glass slides and covers. Having soaked our specimen in water, we lay it on a slide, and by cutting through the stem with one of the needles, close to the attachment of a leaf, we can readily remove the leaf entire, and two or three may be transferred to another slide, and placed in a drop of water under a cover: the same thing may be coughly accomplished by scraping the stem backwards with one of the needles; but in this way the leaves are often torn.

Fig. 36.

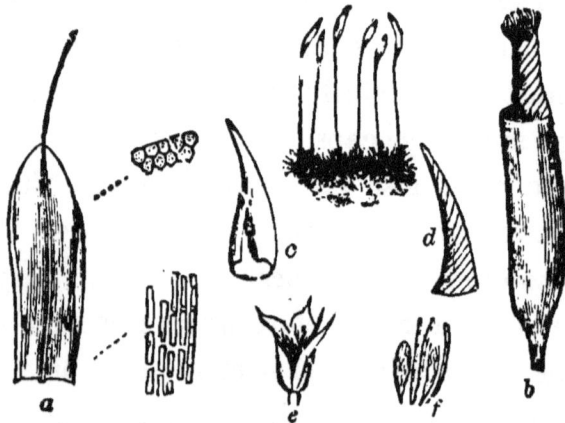

Tortula muralis.

a. Leaf and its areolation. *b.* Capsule. *c.* Calyptra. *d.* Lid.
e. Male flower. *f.* Antheridia and paraphyses.

By examination of a leaf we notice its form, the condition of its margin, whether entire or serrated or bordered; the presence and extent of the nerve; and lastly, and most important of all, the form and condition of its component cells; and for this a higher power is required. With a $\frac{2}{3}$-inch object-glass and C eyepiece we can observe their form, and whether their walls are thickened so as to render them dot-like; their contents, whether chlorophyllose or hyaline; and their surface, whether smooth or covered with papillæ; for often these points are so characteristic, that by them alone we can at once refer a barren specimen to its proper family or genus.

Preservation of Specimens.—This may be discussed under two heads : 1st, as microscopic objects; 2nd, for the herbarium.

1. The parts required for microscopic examination are the capsules and peristome, entire specimens of the smaller species, and detached leaves. The capsules having to be viewed by condensed light, must be mounted dry as opaque objects; and for this purpose I use Piper's wooden slides, with revolving bone cover; and in one of these we may fix a capsule with the lid still attached, another laid on its side, but showing the peristome, and a third with the mouth of the capsule looking upward, a position very useful for the species of *Orthotrichum*, as we are thus enabled to see the inner peristome ; and with them also may be placed the calyptra: should the cost of these be an object, a cheaper substitute may be found in shallow pill-boxes, blackened on the inside.

To preserve the leaves in an expanded state we may employ the fluid media used for vegetable tissues, or, when time is of consequence, Rimmington's glycerine jelly is a convenient material in which to mount them, a ring of dammar cement being first placed on the slide, and within this the liquefied jelly, to which the expanded specimen is quickly transferred, and the cover securely sealed by gold size. Preparations of this kind are of the highest value as types for comparison with actual specimens we may have for determination.

Fig. 37,

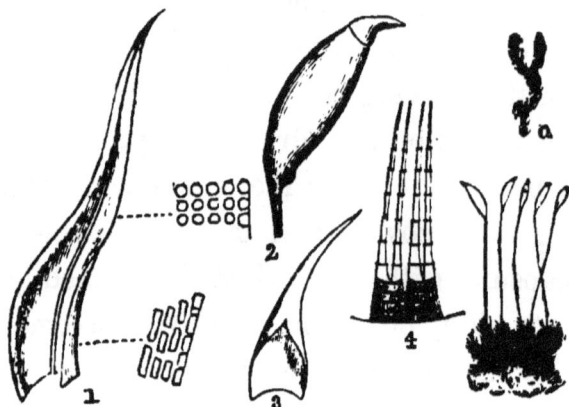

Ceratodon purpureus.

a. Male plant.   1. Leaf and its areolation.   2. Capsule.   3. Calyptra   4. Two teeth of the peristome.

2. In mounting specimens for the herbarium we must be guided by the limits which we have fixed on for the extent of the same; and I may first describe the method adopted for my own collection.

Every species has a separate leaf of cartridge-paper measuring 14½ × 10½ inches, and on this the specimens are fixed, each mounted by a little gum on a piece of toned paper; thus 4 or 6 to 12 specimens, according to size, are attached to each leaf,—varieties have one or more additional leaves; and to each is also fixed a triangular envelope, inclosing loose capsules and leaves for ready transfer to the microscope, and also a label indicating the name, habitat, and date of collection. A pink cover for each genus includes the species, and a stout millboard cover embraces the genera of each family, with the name of which it is labelled outside, the whole shutting up in a cabinet.

Another form is that seen in Rabenhorst's Bryotheca Europæa, quarto volumes of 50 specimens, one occupying each leaf, and so arranged that the specimens do not come opposite to each other. Others again use loose sheets of note-paper, within each of which a single specimen is mounted; but this, from their size, is very cumbersome. Or we may take a single well-chosen typical specimen and arrange many species on a page, as is seen in the beautiful volume of Gardiner's 'British Mosses' or McIvor's 'Hepaticæ Britannicæ.' Whatever plan we adopt, our specimens, once well dried and kept in a dry place, are unchangeable, and are always looked upon with pleasure, each recalling some pleasing associations, or perchance reminding us of some long-lost friend, in companionship with whom they were collected or studied. A stock of duplicates must also be reserved, from which to supply our friends, or exchange with other collectors for desiderata in our own series: these may be kept in square cases of various sizes, cut so as to allow the edges of the top and sides to wrap over the other half folded down on the specimens.

The Hepaticæ of the family Jungermanniaceæ are treated precisely as mosses, the capsules, however, show but little diversity, and will not require separate preservation; but the elaters, or spiral threads accompanying the seeds, are elegant microscopic objects. The Marchantiaceæ must be pressed when fresh, as they do not revive with the same facility as other species, owing to their succulent nature and numerous layers of cells.

Classification.—On this I have fully treated elsewhere ("The Moss World," 'Popular Science Review,' Oct., 1871), and it may suffice here simply to indicate the families of British mosses and their mode of arrangement. The cell-texture of the leaf takes an important place in the characters, and in accordance with this principle the Cleistocarpous or Phascoid group is broken up and distributed in various families.

We have two orders; one indeed, comprising only the genus *Andreæa*, is distinguished by the capsule splitting into four valves united at apex; the other, including the bulk of the species, has in most cases a lid, which separates transversely, and usually discloses a peristome of tooth-like processes. The structure of these teeth again enables us to form three divisions. In the first they consist of a mass of confluent cells; in the second, of tongue-shaped processes, composed of agglutinated filaments; and in the third, of a double layer of cells, transversely articulated to each other, the outer layer composed of two rows of firm coloured cells, the inner of a single series of vesicular hyaline cells, on which the hygroscopic quality of the tooth depends.

Sub-Class SPHAGNINÆ.
Bog Mosses.
Fam. 1.—Sphagnaceæ.

Sub-Class BRYINÆ.
Frondose Mosses.
Order 1.—SCHISTOCARPI.
Fam. 1.—Andreæaceæ.
Order 2.—STEGOCARPI.
Div. 1.—Elasmodontes.
Fam. 2. Georgiaceæ.
Div. 2.—Nematodontes.
Fam. 3.—Buxbaumiaceæ.
Fam. 4.—Polytrichaceæ.

Div. 3.—Arthrodontes.

Subdiv. 1.—Acrocarpici.

*Distichophylla.

| Fam. 5. Schistostegaceæ. | Fam. 6. Fissidentaceæ. |

**Polystichophylla.

Fam. 7. Dicranaceæ.	Fam. 12. Splachnaceæ.
„ 8. Leucobryaceæ.	„ 13. Funariaceæ.
„ 9. Trichostomaceæ.	„ 14. Bryaceæ.
„ 10. Grimmiaceæ.	„ 15. Mniaceæ.
„ 11. Orthotrichaceæ.	„ 16. Bartramiaceæ.

Subdiv. 2.—Pleurocarpici.

Fam. 17. Hookeriaceæ.	Fam. 20. Leskeaceæ.
„ 18. Fontinalaceæ.	„ 21. Hypnaceæ.
„ 19. Neckeraceæ.	

Sub-Class HEPATICINÆ.

Liver Mosses.

Fam. 1. Jungermanniaceæ.	Fam. 3. Anthocerotaceæ.
„ 2. Marchantiaceæ.	„ 4. Ricciaceæ.

Among species which may be generally met with by beginners on the look-out for mosses, we may enumerate the following:

On Walls.—Tortula muralis and revoluta, Bryum capillare and cæspiticium, Grimmia pulvinata, Weisia cirrhata.

In Clay Fields.—Phascum acaulon, Pottia truncatula and Starkeana.

On Waste Ground and Heaths.—Ceratodon purpureus, Funaria hygrometrica, Campylopus turfaceus, Bryum argenteum, nutans, and pallens, Pleuridium subulatum, Dicranella heteromalla and varia, Physcomitrium pyriforme, Pogonatum aloides, Polytrichum

commune, piliferum, and juniperinum, Tortula ungui-
culata and fallax, Bartramia pomiformis, Jungermannia bicuspidata, Lepidozia reptans, Ptilidium
ciliare, Frullania tamarisci.

Shady Banks and Woods.—Catharinea undulata,
Weisia viridula, Tortula subulata, Mnium hornum,
Dicranum scoparium, Hypnum rutabulum, veluti-
num, cupressiforme, prælongum, purum, and mol-
luscum, Plagiothecium denticulatum, Pleurozium
splendens and Schreberi, Hylocomium squarrosum
and triquetrum, Thuyidium tamariscinum, Fissidens
bryoides, Plagiochila asplenioides, Jungermannia
albicans, Lophocolea bidentata.

In Bogs.—Sphagnum cymbifolium and acutifo-
lium, Gymnocybe palustris, Hypnum cuspidatum,
stellatum, aduncum, and fluitans, Jungermannia
inflata.

Rocks and by Streams.—Grimmia apocarpa, Tri-
dontium pellucidum, Hypnum serpens, filicinum,
commutatum, and palustre, Scapania nemorosa,
Metzgeria furcata, Marchantia polymorpha, Pellia
epiphylla, Fegatella conica.

On Trees.—Ulota crispa, Orthotrichum affine and
diaphanum, Cryphæa heteromalla, Homalia tricho-
manoides, Hypnum sericeum, Isothecium myurum,
Frullania dilatata, Radula complanata, Madotheca
platyphylla.

Small as this list is, it will be found to yield ample store for investigation, and if true love for the study be thereby excited, the circle of forms will be found to widen with every new locality visited. If we have contributed in any way to facilitate the pursuit, then is our object fulfilled, and we may conclude with the words of Horace:

Vive, vale ! si quid novisti rectius istis,
Candidus imperti, si non, his utere mecum.

XII.

FUNGI.

BY WORTHINGTON G. SMITH, F.L.S.

WITH the fogs and rains of autumn the fungologist's harvest begins. A few fungi (large and small) appertain to the spring, and some species may be found in every month of the year; but it is not till September has well set in, or October is reached, that the glut of fungi is really upon us. Fungi may generally be met with in abundance for three months of the year; viz. from the latter half of September till the middle of December, the month of October taking pre-eminence for producing the greatest abundance of species. A season of moderate heat and rain is the most productive, for an excessive amount of either dryness or moisture appears to destroy the fecundity of the mycelium, which it must always be remembered is alive and at work (underground) the whole of the year; for, as a matter of course, this year's fungi is produced from last year's spores. These spores are set free in autumn, and at once vegetate and form masses of mycelium, from which next year's crop must spring;

just as the seeds of our wild annuals are self-sown
at the fall of each year and first germinate at that
season. It is a great mistake to suppose that Agarics and Boleti wait till the leaves fall, so that they
may prey upon them; for, as a rule, the larger fungi
never live upon the leaves of the same year as that
in which they (the fungi) come up; fungi live
upon the fallen leaves of the previous autumn. The
spring and summer months will sometimes prove
very productive, especially after stormy weather;
but the collector must always bear in mind that
fungi, like all other things, have their *seasons*. I
have known the fungus harvest quite over by the
end of August, and I have also known it not come
in before December: it depends entirely upon a
certain amount of atmospheric heat and moisture.
A damp summer and stormy August will produce
the crop at the beginning of September; but a dry
autumn, without much rain till November, will delay
the fungus crop till Christmas. Some species appear
regularly *twice* a year from the same mycelium; once
after the rains of March and April, and again in
October. This is the case with *Coprinus atramenta-
rius*, which I have growing (originally from spores)
in a bed of my own garden.

It is useless to go out specially to collect fungi,
either during the dry hot weather of summer or the

frosts of winter; it sometimes, however, happens that odd fungi may be found here and there, in out-of-the-way places, such as the sides of open cellars and sawpits, under bridges, on prostrate logs in streams, in damp outhouses, or about old water-butts, &c. ; therefore I never go out without two or three old seidlitz-powder boxes, some thin paper, and a strong knife, in case any waifs and strays should fall in my way. I have sometimes found good species in a friend's dustbin or cistern, or upon the sides of the open cellar of a public-house. I once found an agaricus on the cornice of London Bridge, to secure which I had to get over the parapet, and was nearly being taken into custody as one tired of life; another time I found a colony of *Coprinus domesticus* upon a friend's scullery wall, and a *Peziza* upon my brother's ceiling. Moral: Fungologists' pockets should, *at all times*, contain one or two small boxes for securing stray and erratic members of the fungus family.

The equipment for a fungus foray differs with the nature of the fungi to be collected. If the plants sought for are wholly microscopic, a small vasculum, knife, pocket-lens, and package of thin paper will be found sufficient; but if Agarics, Boleti, the larger Polyporei, &c., are to be brought home, a more complete set of things will be required, which should

M

include a very small garden-trowel or carpenter's gouge (any saddler or bootmaker will make a suitable leather case for the blades for a shilling or two), a strong knife—such as gardeners use for pruning trees, a few sheets of thin paper, a lens, pocket-compass, and some string. If truffles are desired, a rake is necessary, and the best plan is to carry the iron-toothed end separately in a leather case, and made to screw on to the end of a walking-stick; when not in use, this end can be carried in the pocket with the trowel, &c. It is requisite that the vasculum be large, with straps to carry it over the shoulders; and the collector should be provided with a set of cardboard boxes, large and small, to go inside the vasculum, and to contain the more delicate or choice spoils of the day. Leather gloves and a thin great-coat are good things for the chilly days of early winter—this coat should be provided with at least four large pockets; and, if the weather is inclement, strong boots and waterproof leggings will be found serviceable. An old felt hat and large cotton umbrella are also desirable, for it is only a piece of folly to go into the wet dripping woods with good clothes. As for the umbrella, it should be one of the Mrs. Gamp pattern, of good size, and with a (removable) ring at the end farthest from the handle, so that it may be suspended from branches of trees, &c., whilst the fungi are sorted or examined below, or a frugal luncheon is discussed (perhaps during a passing storm of rain). The string will be found useful for tying up the larger Polyporei; these are frequently of great size, and often weigh many pounds. In collecting, all Agarics should be kept separate as much as possible; for this purpose thin paper, such as is used by stationers and milliners, is indis- pensable; every specimen should be wrapped very lightly in a piece of thin paper before boxing, as the elasticity of the paper not only prevents breaking and bruising, but it also prevents the spores of one species being scattered over another. In carrying fungi about, or sending fresh specimens from one place to another, nothing is so good as this thin paper interspersed here and there with fronds of the common bracken. Sawdust, hay, or wool, should never, on any account, be used: such things totally destroy the plants; but with careful packing with paper and bracken-fronds, fungi may be transported for any distance, by rail or otherwise, perfectly intact and undamaged. In packing the vasculum, see that the heavier plants are at the bottom and the lighter ones at the top; for if packed otherwise any fragile species will be certainly destroyed. I have known a good collection of Agarics rendered worth- less by a loose puff-ball being placed with them,

M 2

which has rolled about with every movement of the collector's body, and damaged big and little species alike, when a piece of paper or a fern-frond or two, to prevent rolling, would have kept all quite safe.

It is hardly necessary to specify localities, because fungi abound everywhere. If leaf fungi are sought for, hedge-sides will produce an abundant crop; if the Agaricini and Polyporei, forests and woods must be ransacked; if the edible species are wanted, rich open pastures (with few exceptions) must be traversed: the various species of truffles must be looked for principally in leafy glades—many prefer a calcareous subsoil, but at times they may be met with even in hedge-sides, town parks, or elsewhere.

When the collection of the day is complete, no species must be allowed to remain in the collecting-cases all night; for if the boxes are not carefully opened and the contents laid out, it will probably be found in the morning that some will have dissolved into an inky fluid, others will have got into the treacle state, whilst a third lot will be overrun with mould, or the smaller ones perhaps entirely eaten up by slugs or larvæ. Few things decompose so rapidly as fungi, especially the full-grown Boleti; these, though apparently perfectly sound one day, will sometimes be a horrible mass of fœtid treacle the next. I have sometimes received large parcels by rail or post when this horrible stinking matter has been dripping out, perhaps all over the carter's hands or down the postman's trousers; for *ladies* always *will* send Boleti in bonnet-boxes, tied with thin twine. Should any extra charge be demanded, on the ground of the insufficiently prepaid postage, or the parcels be unpaid, I invariably refuse to take them in, to the disgust of the parties bringing them. I shall not soon forget an ill-tempered postman who brought me two of these dripping treasures at the same time last autumn, with a demand for extra postage, and his look of silent disrelish as he walked off with one twine-suspended bonnet-box in each hand, the fragrant Boleti-treacle meanwhile manifesting itself upon the pavement. Even when quite fresh, the odour of some species is disgusting in the extreme; for instance, a single specimen of *Phallus impudicus* in the collecting-box will affect a whole railway carriage with the most horrible and sickening stench; whilst the curious truffle *Melanogaster ambiguus* is perhaps worse still, for its abominable odour is perfectly insufferable.

To dry and preserve a collection of fresh fungi is at times a very difficult task; for instance, some species are so entirely covered with a tenacious gluten that if they were at once put between drying-papers, it is certain they would never come out again with the least chance of being recognized by even the most acute fungologist; others are so deli-- quescent that in an hour or two they would dissolve into a watery mass, soak through all the paper, and leave a mere dirty stain between the sheets where the plant was originally placed. As a contrast, some of the Polyporei (as the young state of *Polyporus igniarius*) are so hard that nothing but a steam-hammer would have any chance of flattening them. There is considerable difficulty in ridding the plants from the larvæ with which they are often infested. A few drops of the oil of turpentine will, however, generally drive them from Agarics and other fleshy fungi; and, in regard to the woody Polyporei, a good plan is to place the plants in an oven, or on a hob for a short time, where the heat is not too powerful to destroy the plants, but still sufficiently potent to drive the larvæ from their holes. If this is not done, the collector's experience will probably be the same as mine has more than once been; viz. on opening a package (which should contain some choice dried fungus), to find only a stain, a few skins of dead maggots, and a little dirt—in fact, some of the species in my herbarium, though mostly poisoned with corrosive sublimate, get entirely de-- voured by rapacious and poison-proof larvæ, mites, and minute beetles.

In addition, however, to the mere drying, certain notes and particulars are required, without which the best dried specimens are worthless; and, again, for the larger fungi to be of real service, the spores of each species must be separately preserved. As regards the drying of the fleshy fungi themselves, the process to observe is as follows:—Lay all ordinary Agarics out separately in a dry place, or in a current of dry air from six to twelve, or even twenty-four hours, according to the species, so that they may part with their superfluous moisture, and thus facilitate drying. In the case of species with glutinous pilei, it will be found that the gluten will more or less set, if carefully attended to, in a dry warm place. If the larger fleshy fungi are inadvertently placed under a propagating-glass, or left on a lawn or grassy place, or kept in damp air from over-night till next morning, the chances are that they will never properly dry at all. When the superfluous moisture has evaporated they may be put gently between drying-papers, but the weight put upon them must at first be of the slightest kind; ordinary books, more or less light, will be found quite sufficient; and few, or perhaps no other plants, require such frequent changing as Agarics. An hour, or often less, suffices for the first pressure, when care must be taken to supply them with fresh and per-

fectly dry paper, or they will immediately mould.
It is a good plan, when the plants are half dry, to
take them out of the papers and put them in dry air,
or in a sunny place for a short time (the length of
time being determined by experience and the nature
of the species), to part with more of their moisture :
so, with constant attention and frequent changing of
the papers, very presentable specimens may at last
be obtained. These dried fungi will now be found
very useful for showing the more superficial characters of the plants; but without sections, spores,
and proper notes, they will be next to useless. In
Agarics it is of the first importance to show the
nature of the attachment of the gills to the stem:
and should the stems be furnished with a volva or
annulus, this must be preserved with the greatest
care—young specimens, too, in different stages of
growth, are often of great value. If possible, it is
well to have a series of dried specimens of each
species; one, as in Fig. 38, A, to display the nature
of the tubes in Boletus and the gills in the Agaricini,
whether they are thick or thin, crowded together or
distant from each other, plain or serrated, free or
annexed; another, as at B, to show the pileus,
whether smooth or floccose, plain, warted, or zoned,
and the nature of the margin, whether striate, bullate,
or plain; a third, as at C, to show the attachment of

pileus to stem in infancy; and, fourthly, a section
or thin slice removed from the exact middle of the
young plant from top to bottom, as at D : this will
show the nature of the veil (if present), and whether

Fig. 38.

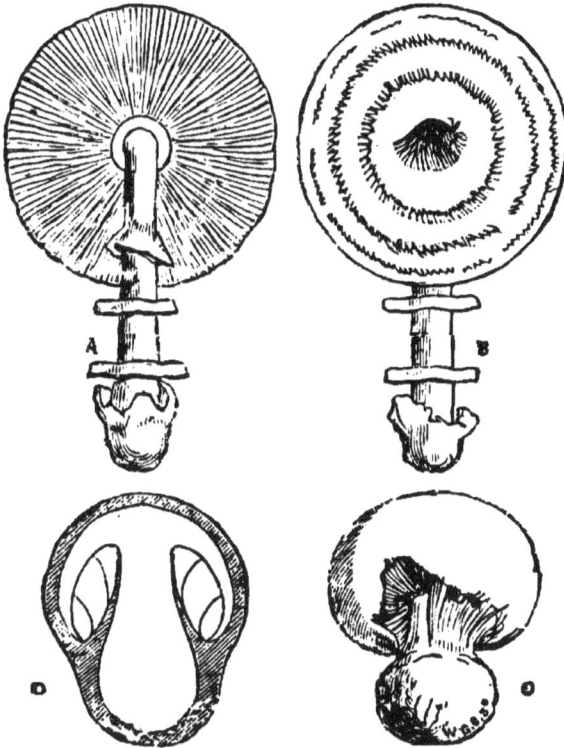

Specimens showing the gills, rings, and stages of growtn.

universal or not; and if absent, whether the margin
is at first straight, incurved, or involute. A similar
section through the mature plant is also required, E

and F (Fig. 39): this will give the attachment of
gills to stems (a character of great importance), and
the nature of the stem itself, whether solid, stuffed,
or hollow.   Great care and experience are required to
cut a thin and perfect slice from the middle of a
tender Agaric or Boletus; for there is often a sort
of articulation at the point G, which causes the slice
to fall in two.   As for preserving fungi in fluids, I

Fig. 39.

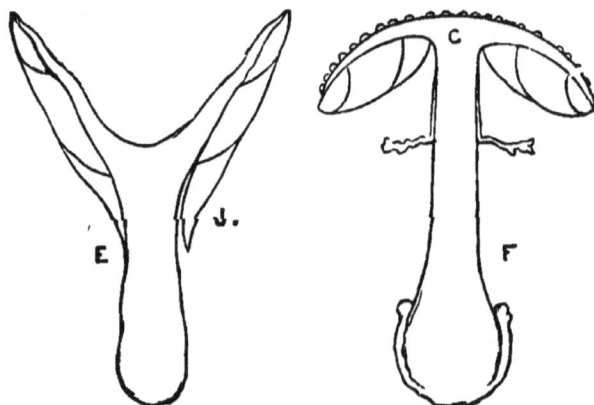

Section cut through Agaricus.

think it in all ways undesirable.   It may more or
less answer for single or unique specimens, or for
large museums, where space is of no consequence ;
but for all purposes of constant reference and private
study, any process of this sort is worthless.   Few
persons, I imagine, would care to have hundreds (or
I might say thousands) of tolerably large glass bottles

of fluids in their houses. It is essential that the
spores should be secured, as their colour and size are
very important. They may be preserved in various
ways: if coloured, they are best kept on white paper,
and if white, on black glazed paper, such as is supplied
to photographers; or they may be at once deposited
and kept on glass slides and covered, or between thin
sheets of mica, such as photographers use. I prefer
the spores free on paper, as they can easily be transferred to glass for examination by breathing on a
corner of a glass slide and just touching it on to the dry
spores; thousands will attach themselves to the glass,
and, moreover, the supply from one fungus appears
to be perfectly inexhaustible. To secure a good batch

Fig. 40.

of spores, it is not sufficient to let the Agaric merely
rest in the position shown at H (Fig. 40), for the
spores will not properly fall when this plan is
adopted; a far better one is to cut a small hole,
about the size of the diameter of the stem of the
fungus, in the centre of the paper on which the spores
are to be deposited: slip the stem through the hole,,

carefully draw up the paper collar, and support the fungus in a small pot, glass, or dry phial (placed under a propagating-glass to keep the plant fresh) as shown in Fig. 41. If it is wished to fix the spores, let the paper be first washed with a thin solution of gum-arabic, which must be allowed to get perfectly dry; the spores may now fall upon the dry gummed paper; and after the deposition the gummed surface must be breathed upon to moisten the gum, and when it has dried for the second time the spores will be fixed, and not readily rubbed off.

Fig. 41.

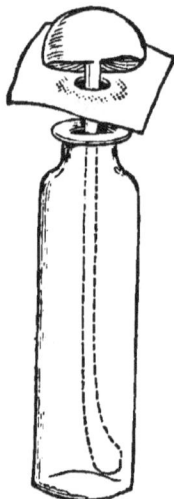

Agaric placed to catch spores.

It is necessary to prepare the woody specimens in a different manner. They must first be perfectly dried before the fire, or in the sun, and then a thin slice must be sawn (or cut with a powerful knife) out of the middle. This slice may be poisoned, as described hereafter, and mounted on the herbarium sheets at once. If the Polyporus is very thin, it may be mounted in company with the slice, but more than one specimen is desirable, as it is indispensable to have both surfaces handy for examination. If the specimens are very large, they are best kept in

wooden boxes, and labelled according to the genera
and sub-genera they contain ; or they may be kept
in drawers, the drawers being divided by partitions
if large, and labelled outside. If boxes are used,.

Fig. 42.

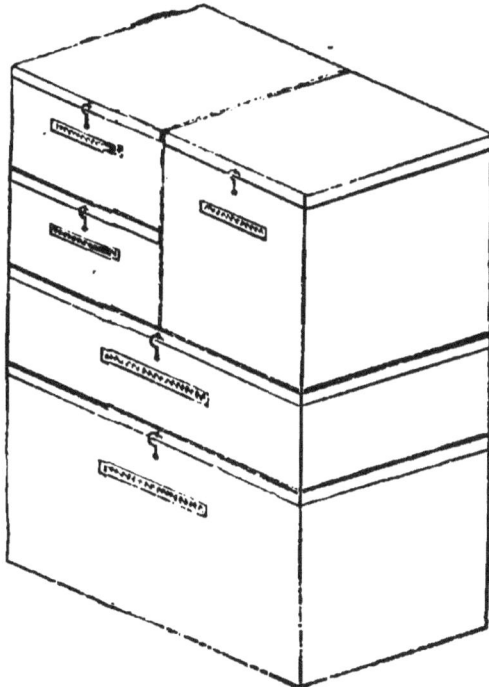

Cabinet for Fungi.

they should all be the same depth ; the height and
width may be doubled or halved according to the
nature of the plants : the plan will be better understood by reference to the diagram, Fig. 42. If this

plan is adopted, there will be no waste space, and the boxes will stand evenly upon a sideboard or against a wall.

Before the specimens are transferred to the herbarium they may or not be poisoned, according to the wish or convenience of the collector. Some of my plants which have never been poisoned remain perfectly uninjured, whilst others, which have been treated with a strong solution of corrosive sublimate, have been devoured by larvæ, &c., introduced, I presume, since the plants were put away. A solution of corrosive sublimate in pyroligneous naphtha, carefully washed over the specimens, has been recommended; but the ordinary poison is oil of turpentine, mixed with finely-powdered sublimate, well shaken before applied. If the specimens are to be glued down, they should be mounted as shown in Figs. 38 and 39, so as to display all their characters, and fixed with poisoned gum tragacanth; but some botanists, and myself amongst the number, prefer to have the specimens *free*. For this purpose I have envelopes gummed to the herbarium sheets, and the specimens (including a small paper containing the spores) are free within the envelopes. Some mites are very fond of the spores of certain fungi, whilst they will not touch the spores of others: therefore, if the specimens are to be kept perfectly intact, gummed paper must be used, or they may be kept between little slips of mica. As to the labelling of the herbarium sheets, I shall not touch upon that, as the plan universally followed is similar to the one used for flowering plants, and described in this volume. Some sub-genera of Agaricus, however (as Tricholoma), are so numerous in species that it will be found requisite to have several wrappers for one sub-genus.

Now as to the necessary notes to be made on the sheets : the points in discriminating fungi differ considerably from those used in naming flowering plants. It is presumed the spores have been preserved by the collector. Now, if he has time, the next best thing is to measure and note them at once, in decimals of an inch and millimetre : a second and essential thing is a note as to the taste of the fungus, whether it is mild, acrid, bitter, &c. This point will be found very useful, as some species are tasteless, insipid, or extremely acrid, bitter, or poisonous : it is only necessary to taste a small piece ; but as so little is really known of the qualities of fungi, unless this is done no advance will be made. I invariably taste every fungus new to me, and have notes to this effect of all the species which have passed through my hands : in some species the effect is very peculiar, sometimes (as in *Agaricus melleus*) it causes a cold sensation at the back of the ears, and swelling of the throat; at others (as in *Marasmius caulicinalis*), the taste proves to be intensely bitter; some are so fiery (as in *Lactarius turpis, blennius,* and *acris*), that the smallest piece placed upon the tongue resembles the contact of a red-hot poker. Often, when 1 have been out botanizing with young men and amateurs, when a dubious *Russula* or *Lactarius* has been shown me to name, I have requested the inquirer to taste it, as, if mild or pungent, the taste might at times decide the species; I have generally found, however, that though certain persons are anxious enough to acquire *names,* they will not burn their tongues to secure them. No fungi that I am acquainted with are really pleasant raw, unless it is *Hydnum gelatinosum,* though many are very good when cooked. A very important thing to note is the odour in the larger fungi : many are very pleasant, like meal; a few are sweet; some resemble stinking fish (as *Agaricus cucumis*) ; one, mice (as *A. incanus*); another, camphor, whilst *Marasmius fœtidus* and *impudicus* are like putrid carrion ; others are like burnt flannel, garlic, rotten beans, and almost every imaginable disagreeable thing. The habitat is of great importance : if the plant grows upon trees, the tree should be named ; or if parasitic upon any other material, the matrix should be named with the place. The viscidity, dryness, or bibulosity must be given, and in the *Agaricini,* any notes that may suggest themselves as to the presence or absence of a veil, volva, or trama, and whether the gills have a habit of separating from the stem, as at J (Fig. 39), must be carefully noted.

The study of the larger fungi has been to me one of the greatest pleasures of my life: when all things else have failed, this has never failed; it has taken me into the pleasantest of places and amongst the best of people. Had it not been for fungi, I should have been dead years ago; often tired, jaded, and harassed with business matters, a stroll in the rich autumn woods has given me a renewed lease of life. In these favourite haunts I never tire or flag; rain, fog, and mud, never detract from the pleasures of the woods to me—I am only depressed in the hot, dry weather of midsummer. In the autumn I constantly visit the forests, with all my collecting paraphernalia; I sometimes take a saw to cut off the big, woody, fungus excrescences of trees. I was once fortunate enough to find a ladder in a wood, which proved invaluable for ascending the beeches in search of *Agaricus mucidus,* &c. I, however, find fungi everywhere : I only go round the corner, and there they are. I often visit a neighbouring builder's yard, and descend the sawpits, to the amazement of

N

the operatives: some of the rarest species of our Flora, and many new ones, I have found within a few minutes' walk of my own house. I once found a rare *Lentinus* on a log as it was being carted down King William Street, and a year or so ago an undescribed *Peziza* flourished inside my cistern.

Collecting fungi is not without its humours as well as its pleasures, as the following will show. I once saw a portly, well-dressed gentleman walking along the high road, with his vasculum over his shoulders, and carrying home (one in each hand) a pair of cast-off, rotten boots, discarded by some vagrant; the rotting leather having produced a crop of rare microscopic fungi. At times abominable cast-off fœtid gipsy rags will be lovingly taken from out a ditch, and choice pieces cut out and consigned to the vasculum of the cryptogamic botanist; at other times some rare species will be seen "up a tree," and it has several times happened in my presence that one enthusiastic botanist has got on to the shoulders of another to secure a prize, or even waded into a pond to get at some prostrate fungus-bearing log. The humours of truffle hunting are manifold. I have seen a gentleman trespass, on hands and knees, through a holly hedge, on to a gentleman's lawn, and there dig up the turf in some promising spot, risking an attack from the house-dog, or a few

shots from the proprietor; the said trespasser meanwhile armed with a rake, gouge, and dangerous-
looking open knife. Country labourers are often
sorely puzzled by the acts of cryptogamic botanists;
they stand agape in utter amazement to witness
poisonous "frog-stools" bagged by the score. Oft-
times one gets warned that the plants are "deadly
pisin"; but collectors are usually looked upon as
harmless lunatics, a climax in this direction generally being reached if a gentleman in search of
Ascoboli and the dung-borne *Pezizæ*, sits down, and
after making a promising collection of horse or
cow-dung, carefully wraps these treasures in tissue
paper, and puts them in his "sandwich-box."

One word of warning to the beginner—never, on
any account, amass and put away a lot of imperfect
materials with insufficient notes, for in the end they
will prove worse than useless. To name fungi with
certainty the fullest notes and most complete materials are indispensable: without these nothing whatever can be done. It is far better to laboriously
make out twenty species, and know them in all their
aspects for certain, than to amass imperfect materials
of two thousand without any sound botanical knowledge. If the former course is pursued, the study of
fungi will prove a never-failing source of pleasure to
the mind and of health to the body.

N 2

In conclusion, 1 cannot do better than quote a few words written by the illustrious Fries (now more than eighty years of age) in the preface to a recent work of his on Fungi. He says: "Now in the evening of my life, I rejoice to call to mind the abundant pleasures which my study of the more perfect fungi, sustained for more than half a century, has throughout this long time afforded me. . . . . . Therefore, to botanists, who can wander at will the country side, I commend the study of these plants as a perennial fountain of delight and admiration for that Supreme Wisdom which reigns over universal nature."

XIII.

LICHENS.

By the Rev. Jas. Crombie, F.L.S., etc.

Much as it is to be regretted, it cannot be ques-
tioned that of those who have devoted themselves to
the study of botany, lichenists have always been
"few and far between." While flowering plants
have had their hosts of enthusiastic students, and
while other classes of cryptogamics have had due
attention paid to them, the study of lichens has,
up even to the present time, been but too much
neglected. To many indeed the term conveys only
some faint and confused idea, and though they know
that there are plants so called, they are at the same
time utterly ignorant of their nature. With flowering plants, ferns, mosses, seaweeds, and even fungi,
they have at least some acquaintance, more or less
accurate; but lichens they generally pass by with
indifference, regarding them merely as "time-stains"
on the trees, the walls, and the rocks where they
grow. Nay, we have even met with some professed,
and otherwise well-informed botanists, who, while
recognizing certain of the larger and more con-

spicuous species as lichens, yet fancied that many of the smaller and more obscure species were merely inorganic discolorations. It is certainly very difficult to account for such a state of matters at the present day, when so much attention is being paid to almost every other class of plants. Vainly have I sought either in the nature of the case itself or in my conversations with botanists, for any intelligible solution of such apathy and neglect: though many good and sufficient reasons have presented themselves to my mind why they should be regarded in a very different light. It cannot with any show of propriety be objected that lichens are an uninteresting class of plants, and consequently undeserving of serious study. So far from this, they are in various respects as interesting not only as any other class of cryptogamics, but also as many other plants, which occupy a higher and more conspicuous place in the scale of vegetation. Being as it were the pioneers of all other plant life, for which they serve to prepare the soil on the coral islet and the barren rock,—constituting the most generally diffused class of terrestrial plants on the surface of the globe, from arctic lands to tropical climes,—presenting essential simplicity of structure, being composed entirely of an aggregation of cells, though at the same time this is amply compensated for by endless variety of form,

—adorning as they do, with their variously coloured thalli and apothecia, the most romantic and the most dreary situations,—affording in some cases valuable material for the dyer and the perfumer, nay, even for medicinal purposes,—supplying, as some of them do, more or less, nutritious food for man and beast, under circumstances and in regions where no other can be had,—it is very evident that the prevailing neglect of them cannot arise from their being in any way uninteresting, and destitute either of beauty or utility. Nor does this, as might be inferred, result from any peculiar difficulty attending their study. There indeed seems to be a notion prevalent, not only amongst the students of phænogamic, but also amongst those of cryptogamic plants, that there are, somehow or other, almost insuperable difficulties connected with the pursuit of Lichenology. Now, it is quite true that the correct study of these plants is by no means an easy one, and that an accurate knowledge of them is not to be obtained in a day or an hour; but the same may, with equal truth, be said of any other branch of Phytology, which requires minute research and microscopical examination. Here, as elsewhere, there is no royal road to learning, and the difficulties which lie in the way must be boldly faced. If the student can only muster up sufficient courage to cross the threshold and prosecute his investigations with zeal and steady perseverance, he will find in this, as in other cases, that the difficulties which looked so formidable at a distance, will, one by one, be successfully surmounted.

But to whatever cause the paucity of lichenists, both in our own and other countries, is to be attributed, it certainly does not originate in any difficulty connected with their collection and preservation. In fact, there is no other class of plants, where these, and more especially the latter, can be so easily effected, at a little expenditure of time and trouble. A few simple directions are, therefore, all that are necessary to be given on these points. As to the collecting of lichens, it has already been intimated that they are almost universally distributed, though of course in this respect subject to the same laws as the higher orders of vegetation. In our own country we have now a list of about eight hundred species, constituting by far the greater proportion of the Lichen Flora of Europe. In most parts of Great Britain and Ireland, a very fair number of these may readily be gathered, capable, as they are, of existing in almost every situation where they can derive requisite nourishment from the atmosphere. On the rocks and boulders of the seashore and the mountain-side, on the trunks and branches of trees in woods and forests, on peaty

soil of bare moorlands, and on stone fences in up-
land tracts, nay, even on old pales and walls in
suburban districts, a goodly harvest may generally
be reaped. Few localities indeed there are, within
the area of these islands (London and its environs,
where the atmosphere is so impregnated with
smoke, being the chief exception), in which the
lichenist will find his occupation gone. True, it is
only in some more favoured tracts, chiefly maritime
and montane, that he can expect to meet with many
of our rarer species; but even in most lowland
districts, especially such as are well wooded, he
may, with profit, pursue his researches, and collect
various of the more common species. These will
just be as useful in making him acquainted with the
structure and physiology of lichens as though he
had gathered the rarest that grow on Ben Lawers
or by Killarney's lake. The apparatus requisite for
collecting is neither complicated nor expensive.

A tin japanned vasculum, or what is perhaps
better still, a black leather haversack, of larger or
smaller dimensions as the case may be, suspended
over the shoulder by a strap, is of course indispensable for holding the specimens gathered. The
latter of these we have found to be more generally
convenient, as we can take it with us also for a short
ramble, without its attracting so much attention

from curious rustics, as the less-known and more singular-looking vasculum. Two sets of instruments are also necessary for removing the plant from the substratum on which it grows, as well as for breaking off in many cases a thin portion of the latter along therewith. These are a geologist's hammer and chisel for such as grow on rocks, boulders, and stones; a gardener's pruning-knife for such as grow on trees, pales, and the ground; as also an ordinary table-knife for detaching, by insertion under them, such foliaceous species as can thus be separated from the substratum. To these must be added several sheets of soft and moderately thick paper, cut into different sizes (some newspapers suit remarkably well), in which to wrap up the individual specimens and prevent them rubbing against each other; a few card-boxes also, of various sizes, in which for greater safety to place any of the more brittle species, or fragments of the rarer ones, by themselves; and a pocket-lens of good magnifying power, by which we may be able to detect on the spot those minuter species which the naked eye can with difficulty distinguish. With these the lichenist is fully equipped for an excursion, whether " near at hand or far away," and, with waterproof and umbrella, is ready to take the field even in threatening weather. A good deal of discrimination must be used in the selection of specimens for removal, which, in all cases where such can be obtained, ought to be fertile, with both apothecia and spermagones fully developed. Hence, such as are too old or too young, may be passed by, as neither the spores nor spermatia, by which alone, in many instances, they can be determined, will be found in a normal condition, any more than the thallus itself. The specimens gathered ought in every case to be of sufficient size to show distinctly the character of the thallus and of the fructification. Where, however, the thallus, as it frequently does, spreads very extensively over the substratum, it will be sufficient to break off such a portion from the circumference towards the centre, as will give an adequate idea of the more important characteristics of the plant. This is a point of considerable consequence; for should a portion be taken off from the circumference alone, or from the centre alone, it will often be entirely unsuitable for showing the real nature of the plant, and be quite useless for purposes of description. A little experience, however, will serve to prevent the commission of a mistake, into which, judging from the number of imperfect specimens which are sent me to be named, beginners are very apt to fall. Practice will also in time enable the tyro to use the hammer

and chisel in such a way as to obtain neat specimens
of saxicole species—a matter of importance with
respect to their subsequent mounting.   As to the
best season for collecting, I need scarcely remind
the reader that lichens are perennial plants, re-
markable for their longevity, and that during the
whole year round they may be found in fruit.   The
lichenist has not to wait for any particular month
or months, as other botanists have to do, before he
can collect the objects of his search in a fully-
developed condition.   Spring, summer, autumn, and
even winter, except when the snow conceals all
vegetation beneath its white mantle, are all alike to
him, and in each he will find every species of lichen
in perfection.   At the same time, he will be most
successful after a shower of rain or a slight frost
has fallen, inasmuch as, becoming swollen with the
moisture then imbibed, many of the minuter species
which might otherwise be overlooked, are more
readily perceived, and the foliaceous species more
easily removed from the substratum to which they
are more or less closely affixed.

Nothing more need be said on the collecting of
lichens, as a short experience will be more useful
than further details.   We proceed, therefore, to give
a few hints on their subsequent preservation.   This
is a very easy process, presenting no difficulty what-

ever, and occupying but little time. We shall
suppose that the collector has returned from a
successful expedition, with his vasculum or haversack well filled with specimens from all sorts of
habitats. Opening the papers in which they have
been wrapped up, he will take them out one by one,
and place them separately upon a table, over which a
newspaper has previously been spread. If gathered
in wet weather, they ought not to be left long in the
papers, as in this case they are very apt to become
covered with mould. After allowing them to remain
in this position till they are thoroughly dry, he may
at once proceed with hammer and chisel, or with
knife and scissors, to reduce to a suitable size such
of them as he could not conveniently thus manipulate
in the field. When this is done, they may then be
affixed with gum, of a rather thick consistency, to
slips of white paper, with the locality and date of
their collection written beneath. There will be no
difficulty felt in thus affixing saxicole, corticole, and
lignicole species, though where the nature of the
stone or wood is more absorbent, several applica-
tions of the gum may be necessary before they
properly adhere. With terricole species, however,
a somewhat more lengthened process is necessary,
owing to the brittle nature of the substratum, in
consequence of which, if not properly preserved,

they often crumble into dust in the herbarium. To prevent this, M. Norman, of Trömsoe, Norway, has recently prescribed a solution of isinglass in spirits of wine, which, when liquefied in a vessel plunged in water of the temperature of 25°–30° C., is greedily imbibed by the earth, and becomes inspissated into a solid gelatine at a temperature below 15°. This solution may be applied until the earth becomes thoroughly saturated, and after it is perfectly dry, the specimens will possess sufficient hardness and tenacity, and may then be mounted like the others. So far, however, as my own experience goes, I have found a weak solution of gum-arabic, frequently repeated, and applied to the under surface and edges of the specimens, to be quite as efficacious; and if after becoming thoroughly dry, they be first affixed by a thicker solution to slips of thin tissue-paper, they will be equally ready for being mounted as above. Either of these two methods may also with advantage be applied to such species as grow upon decayed mosses. Slight pressure may be applied to the thallus of fruticulose, filamentose, and foliaceous species, in order that they may lie better in the herbarium; but this should be done only to a very limited degree, so as not to obliterate the normal appearance of the branches or lobes. As the character of the under surface of the thallus is frequently of great importance, at least in foliaceous and fruticulose plants, a portion of this, not necessarily detached, should be turned over, for facility of inspection, and pressed down on the paper, before the specimens have become quite dry and rigid. In order to destroy any insects that may be upon the plants when gathered, or by which they may afterwards be infested, lichenists at one time were in the habit of poisoning them with corrosive sublimate. Frequent exposure, however, to the air in dry weather, and the presence of a little camphor, will be quite sufficient to prevent any mischief from this source.

But having thus arranged, though the arrangement is but temporary, the specimens gathered, on slips of white paper, the next and most important point is their due examination and determination. This, in the present advanced state of Lichenology, is unquestionably, in many cases, a task of considerable difficulty, and in the short space at our disposal it would be quite impossible to give anything like an adequate explanation of the mode in which this is to be effected. Suffice it at present to say that sections must be made of the thallus to ascertain the character of its different layers, as also sections of the apothecia and spermagones to ascertain the nature of the spores and spermatia. For both purposes a good microscope, with $\frac{1}{4}$-inch object-glass, is absolutely indispensable to the student. The examination of the spores, upon which, in so many cases, the determination of the species chiefly depends, should present little or no difficulty, at least to the fungologist. It may be readily effected by moistening the apothecium with water, and then, with a dissecting-knife, making a thin vertical section through its centre. Putting this on a glass slide, or in a compressorium, in a drop of hydrate of potash, and

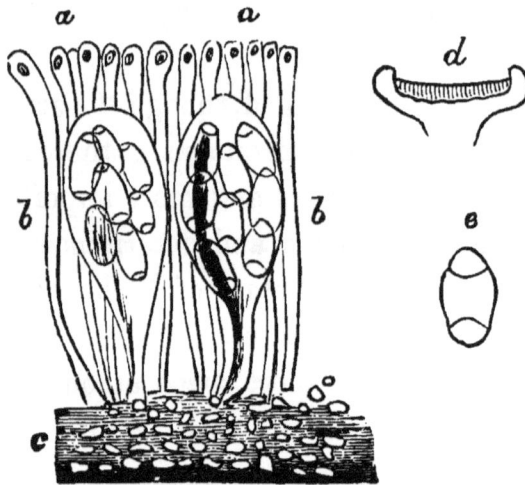

Fig. 43.

Section of *Physcia parietina.*
a. Paraphyses.   *b.* Asci with spores.   *c.* Hypothecium.   *d.* Section of
apothecium.   *e.* Spore.

then placing it under the microscope, a view will be obtained of the asci, spores, paraphyses, hypothecium, &c., each of which may afterwards be insulated and examined more minutely in detail. Take, for example, the well-known beautiful yellow lichen (*Physcia parietina*), so common everywhere on walls, rocks, and trees, and treat a very thin section of the mature apothecium as before mentioned. Under the microscope it will appear as represented in Fig. 43.

In the same way the spermagones may be examined, when the nature of the sterigmata and spermatia will be apparent. By cutting across the thallus of the above species, we can perceive even by the naked eye that it consists of three different layers, which when microscopically examined present the appearance shown in the above figure.

Fig. 44.

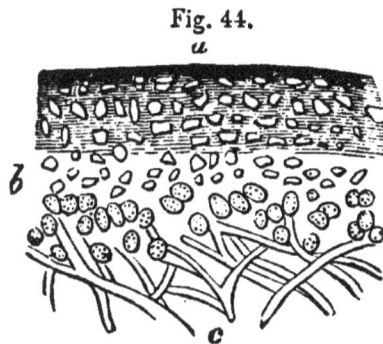

Section of *Physcia parietina.*
a. Cortical stratum. b. Gonidic stratum. c. Medullary stratum.

But in addition to this microscopical examination, it is also requisite to observe the different chemical reactions produced on the asci or the hymeneal gelatine with iodine (I), which will tinge these either bluish or reddish wine-coloured, or else leave them uncoloured. Similarly the thallus, including both the cortical layer and the medulla, may be tested with hydrate of potash (K), and hypochlorite

of lime (C), the latter being applied either by itself
or added to K when wet. In some cases no reaction
will be produced by these either upon the cortical
stratum or the medulla; in others they will be
tinged yellowish or reddish. The formulæ for the
preparation of these reagents are: for iodine, iodine,
gr. j; iodide of potash, gr. iij, distilled water, ½ oz.;
for hydrate of potash, equal weights of caustic potash
and water; for hydrochlorite of lime, chloride of
lime and water of any strength. After correctly
ascertaining the specific name of the specimens
collected, this is to be written on the slips of paper
to which they are affixed, above the locality and
date, and the best of them, including all varieties
and forms, selected for subsequent mounting in the
herbarium. This may be effected either in the same
way as the mounting of phanerogamic plants, or by
affixing the specimens to pieces of millboard covered
with white paper, and arranging them according to
the order of the genera and species in the system of
classification which may be adopted. For facility
of reference the latter is undoubtedly the preferable
method; and if the cards are disposed in a cabinet
with shallow drawers, they will not, so far at least
as our British species are concerned, be found to
occupy too much space.

SEAWEEDS.

By W. H. GRATTANN.

In some articles published in 'Science-Gossip' a few years ago, I gave some directions for collecting and preserving Marine Algæ, or seaweeds, and although, I think, it will be difficult to simplify those directions, or even to add much that would be really serviceable to young beginners in this delightful pursuit, it is my intention, in going over the ground once more, to be as explicit as I possibly can ; and here, on the threshold of the subject, I have a few words to say to one or two occasional contributors to that journal, who, in calling attention to the beauty of marine vegetation, and urging young persons to collect and preserve Algæ, have advised them to ignore books on the subject, and go to the shore, use their own eyes, and collect for themselves, &c. I am sorry very greatly to differ from such advice. Collecting in this way may be amusing enough to those who care not for science, but when it leads to parcels of seaweeds, picked up at random, being sent to botanists with a request that the names of such plants should be sent to the writer, it is the reverse of pleasure to the scientific botanist, for it gives him infinite trouble, and enables him to convey but very imperfect information to his applicant. The editor of that journal has often been thus appealed to, and packages of decayed rubbish have frequently been sent to me for examination, containing species or rather fragments of plants, which, for the most part, were utterly worthless and defied identification.

Almost all collectors commence by mounting plants which a little experience proves to be really what the old poet termed "*alga projecta vilior*"; but as seaweed-gathering, like everything else, requires practice, beginners must not be disappointed because they do not find rarities or fine specimens whenever and wherever they may seek for them.

When I think of the difficulties I experienced at the outset of my study of marine botany, especially in the collecting and drying of seaweeds, I feel strongly inclined to urge all beginners to obtain some information concerning Marine Algæ before they go to the shore to collect for themselves. A very few hours of study with an experienced algologist, or even a perusal of some illustrated work on British algæ, will save much trouble and materially assist the unpractised eye in selecting specimens for the herbarium. I may here mention as highly useful to incipient algologists Dr. Landsborough's 'British

Seaweeds,' and Professor Harvey's 'Manual,' either
of which may be obtained for a few shillings; but if
my readers are resident in London, I advise them
to pay a few visits to the Library of the British
Museum, and there inspect Dr. Harvey's 'Phycologia
Britannica.' In this magnificent work they will find
coloured figures of nearly every British seaweed,
with drawings from magnified portions, and various
structural details of the highest value to students;
and I once more impress on all collectors the importance of some degree of book-learning ere they
sally forth, bag or vasculum in hand, to cull the
lovely "flowers of the ocean," or gather what best may
please them from the rejectamenta on the shore.

If the collector wishes to learn, not merely the
names of plants, but to distinguish *species*, he will do
well to provide himself with a copy of Harvey's
little volume the 'Synopsis of British Seaweeds,' and
a Stanhope or Coddington lens, by means of which
he can examine portions of delicate plants as he
finds them, and compare them with the descriptions
given in the 'Synopsis'; in this way, if he have any
success during his excursions, he will quickly become
familiar with most of the plants which are cast ashore
or grow within tide-marks.

Time will not admit of, neither is space at present available for, a single line beyond what may
be practically serviceable to my youthful readers;

therefore I will hasten to describe the course of action in seaweed-collecting as I have practised it for many years. At once, then, to the shore, but not to the sandy shore, for only useless decayed rubbish, or here and there some straggling plants of *Zostera marina*, or grass-wrack, will be met with there. The collector must away to the rocks, and search carefully every pool he meets with, from a little distance below high-water mark, and so on down to the water's edge, always remembering that it is better to collect while the tide is receding than as it is coming in.

Presuming that few persons will think of collecting seaweeds much earlier than the month of May, let me observe that most of the accessible species of olive and green plants which grow on rocky shores and in tide-pools, will be found from May to June in pretty fair condition, but very few red plants, except those which grow on the shady sides of rock-pools, or under the shelter of the larger olive weeds, will be met with until a considerable space is laid bare by the receding water at the low spring tides, about a day or two before and after the full moon.

As nearly all the *rare* red weeds grow in deep water, they are seldom taken in any degree of perfection unless they are dredged; but in the summer months, say from June to the end of August, many fine plants are occasionally thrown up from deep

water, and others are found growing on the stems
of the great oar-weeds, portions of which are cast
ashore, beautifully fringed with one or more species
of Delesseria and other rare Rhodosperms—in fact,
during the rising tide, diligent collectors may secure
many a lovely deep-water plant as it comes floating
in, but which, if allowed to remain long exposed to
the action of sunlight, will fade in colour and decompose before it can be mounted. This is especially the case with, all the soft gelatinous red plants,
such as the Callithamnia, and all the Gloiocladiæ, as
well as a few of the softer olive weeds; and here I
may observe that there is one genus of beautiful
olive plants, the *Sporochnaceæ*, which must on no
account be put into the vasculum with any of the
delicate red plants, for they not only very rapidly
decompose, but injure almost all others with which
they are placed in contact. The species are not
numerous, and they may be easily recognized, after
having been previously studied from the coloured
figures either in Harvey's 'Phycologia,' or in Brad-
bury and Evans's 'Nature-printed Seaweeds.' It is
also a curious fact respecting this genus, that while
they are all of a beautiful olive tint in the growing
state, they invariably change to a fine verdigris-green
in drying; and indeed this is very generally the case
with the filamentous olive weeds, the Fuci, or common
rock-weeds, as constantly turning quite black after

mounting: whence the term, that of "Melanosperm," which is given to the subdivision to which all the olive weeds belong.

As there are so few seaweeds which have generally known common names, I shall make no apology for using the names by which they are known to science, presuming that all intending collectors will, as I have already suggested, gain *some* knowledge of Terminology ere they go out "seaweeding."

Beginners should be cautioned against the very natural error of bringing home too many plants at a time ; they must be moderate in their gatherings, or be content to risk the loss of some choice specimens, which will decompose unless they are attended to before night. The first thing to be done upon arriving at home, is to empty the collecting-bag into a white basin of sea-water, and to select the best and cleanest plants as soon as possible, giving each a good swill before placing it in another vessel of clean water, and getting rid of rejected plants at once, so that the basin first used will be available for re-washing the weeds before they are severally placed in the mounting dish. When a day is fixed on for seaweeding, the collector should order a large bucket of clean sea-water, which, after being left to settle, should be strained through a towel, so as to be as free as possible from sand and dirt. Two or three large pie-dishes will be necessary, the deeper the

better, and white, if such can be obtained. Place •
these on a separate table with towels under them,
and reserve a table specially for the mounting dish
and the parcels of papers, calicoes, and blotting-
papers. The large white bath used in photography
is very well adapted for mounting seaweeds; the lip
at one corner is convenient for pouring off soiled
water, and its form—that of an oblong—is most
suitable for receiving the papers on which the
plants are to be mounted. Beside this vessel should
be placed the following implements—a porcupine
quill, two camel-hair pencils (one small, the other
large and flat), a pair of strong brass forceps, a penknife, a pair of scissors, a small sponge, an ivory
paper-knife, and two thin plates of perforated zinc
somewhat less in length and breadth than the inside
of the mounting dish.

Smooth drawing paper, or fine white cartridge
paper, is generally employed for mounting. The
operator should be provided with three different
sizes of paper, and these should have each a piece
of very fine calico and four pieces of blotting-paper
to correspond. The process of mounting one of the
filamentous or branching species is as follows:—The
specimen being cleaned and placed in the mounting
dish, a piece of paper of suitable size is laid on one
of the perforated zinc plates, and both are then
slipped quickly under the floating weed. The root

or base of the specimen is then pressed down on the
paper with a finger of the left hand, while the right
hand is employing the forceps or porcupine quill in
arranging the plant in as natural a position as
possible, ere the zinc plate is gently and gradually
raised at the top or bottom, as may be necessary, to
ensure a perfect display of every portion of the
plant; but if, upon drawing it out of the water, it
should present an unsightly appearance from too
thick an overlapping of the branches, the whole must
be reimmersed, and a little pruning of superfluous
portions may be employed with advantage to the
specimen and satisfaction to the operator. Care
should be taken that the water be drained off the
paper as completely as possible before the calico is
laid over the plant, and this is accomplished by
raising the paper containing the plant as it still lies
on the zinc plate, and transferring it to a thin board
placed in an inclined position against one of the
basins, and with the large camel-hair pencil *paint*
off the water as it runs away from the specimen, and
absorb what remains, when the paper is laid flat,
with the sponge. Delicate species may be left to
drain for a few minutes, while the operator is arranging other specimens. When the water is sufficiently
drained off, the paper is then laid on the blotter, and
the piece of calico is placed upon the plant—a sheet
of blotter being laid upon the calico.

Care should be observed in subjecting plants to pressure, which, in the first instance, should be sufficient only to help the absorption of water. The first set of blotting-papers should be changed in half an hour after the whole batch of specimens have been placed in the press, and these must be thoroughly dried before they are used again. After the second or third change of blotters, the plants should remain under strong pressure for two or three days; but the pieces of calico must not be removed until it is pretty certain that the papers and plants are quite dry.

With the exception of the Fuci or common rock-weeds, I never place seaweeds in *fresh* water: with these, especially *Fucus serratus, F. nodosus, F. vesiculosus,* and *F. canaliculatus,* a few hours' immersion in fresh water is an advantage, as it soaks the salt out of their fronds and renders them more pliable. As all the Fuci turn black in drying, and few of them adhere well to paper, I arrange my specimens in single layers between the folds of a clean dry towel, and keep them under pressure until they are quite dry; they may then be put away loosely, or gummed on sheets of paper.

The foregoing directions for mounting filamentous seaweeds are applicable to all the branching species of Olive, Red, and Green plants; but in each of the three subdivisions there are a few species which are so gelatinous—in fact, so soft and spongy, that they require the utmost care during pressure, otherwise they adhere to the calico and break off in fragments as it is drawn away. Such plants must be left to dry in a horizontal position for an hour or so before the calico and blotters are placed over them, and pressure must be very slight until they have adhered closely to the paper. Among the Chlorosperms, or green plants, there are the various species of Codium, young plants of which only are manageable or indeed desirable. In the Melanosperms, some species of the genus *Mesogloia* will require care and patience in mounting, as well as the long string-like plant, known as *Chorda filum*; and again, the spreading tuberous mass called *Leathsia tuberiformis*, portions of which should be cut from the rock, the sand scraped and washed out, then laid on the wet paper, and allowed to shrink for some hours ere calico blotters and pressure be applied. These difficulties are much more numerous among the Rhodosperms, or red seaweeds, experience only teaching the best method of treatment. I will, however, mention the names of some very troublesome plants, the fronds of which, if subjected to pressure too soon, burst and discharge their carmine contents; not only presenting an unsightly appearance, but destroying the specimen. These are *Griffithsia corallina*, *Dudresnaia coccinea*, *Naccaria Wigghii*, all the *Chylocladia*, and the rare *Gloiosiphonia*, as well as the slimy worm-like plant known as *Nemalion multifidum*.

In addition to these troubles among the red plants, there is an opposite difficulty connected with several Rhodosperms which must be pointed out; and that is owing to an absence or scarcity of gelatine in their substance, which is in some of a stout, leathery, or horny nature, and in others is due to a coating of carbonate of lime, which completely envelops the vegetable structure. Among the former may be mentioned the several species of Phyllophora, and several among the genera Gigartina, Chondrus, and Sphærococcus; and in the latter, all the calcareous Algæ, especially the well-known *Corallina officinalis* and *Jania rubens*. All these, and several others of a membraneous nature, among the olive as well as the red weeds, must be first mounted in the ordinary manner, and when they are tolerably dry and begin to shrink away from the paper, fill the mounting-dish with stale skimmed milk; refloat the plants on their papers in the milk, and indeed go through the same process as before with the sea-water, but be careful to absorb all the milk from off the surface of the plants and the back of the papers, and then, after the usual time for drying and pressing, the most obstinate seaweed will be found adhering perfectly to the paper, and will remain so permanently.

One more difficulty must be referred to for the

benefit of young beginners, who, in mounting some of the Laminaria and that peculiar olive weed called *Himanthalia lorea,* may wish to preserve the thick-branching roots and stems. First wash the roots as clean as possible, and then, with a sharp penknife, make a clean cutting horizontally of the whole root and some little distance up the thick round stem; then, after having removed the cut portions, place the inner surface of the root and stem on the paper, and the gelatinous matter which oozes from the plant will cause the roots to adhere firmly to the paper, and in drying, the usual olive tint of the various species of Laminaria will be finely preserved. Some botanists employ a mixture made of isinglass, dissolved in alcohol, to fix some of the horny or robust species on paper; but if gum be made use of, it is better to employ gum tragacanth than gum-arabic, because, in drying, the former has none of that objectionable glare which is peculiar to gum-arabic.

As regards the best method of pressing seaweeds, I think I can hardly do better than refer my readers to the figure of a Seaweed Press (Fig. 45), which I invented for myself many years ago, in which I have pressed many thousands of beautiful seaweeds. Almost any degree of pressure can be obtained in it: first, by the thumb-screws on the iron rods at each corner, and, finally, by means of the clamp which is strapped on the top of the press. Any intelligent cabinet-maker or ironmonger could provide such a press from an inspection of the figure; the cost, of course, varying with the dimensions and the number of boards.

Fig. 45.

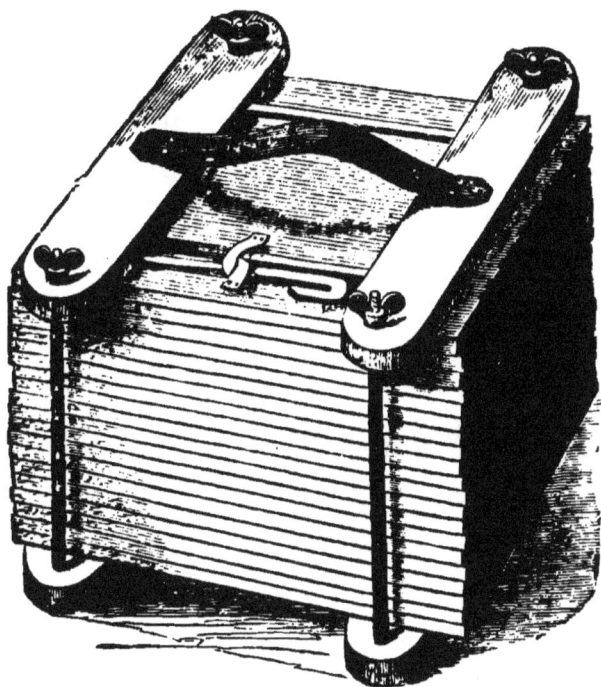

Seaweed Press.

With respect to localities favourable to seaweed-gathering, I may specially mention the south coast of Devon; from Exmouth, where *Bryopsis* and *Padina pavonia* grow in perfection, to Torquay and the coves of Torbay, and down the coast to Plymouth,

Cawsand Bay, and finally Whitsand Bay, the "happy hunting-grounds" of the enthusiastic algologist. On the north-east coast, Filey and Whitby must be mentioned, as well as the shores upwards from Tyne-mouth to Whitley. Peterhead is also a good locality, the rare *Ectocarpus Mertensii*, *Odonthalia dentata*, and *Callithamnion floccosum* being found there in abundance. Other favourable stations in Scotland, well known to me, are Lamlash Bay and Whiting Bay; nor must the Isle of Wight be forgotten, for in the rock-pools, at Shanklin especially, the most magnificent form of *Padina pavonia* may be found growing during the summer months in the utmost profusion.

In conclusion, I beg leave to inform my readers that I have recently published a volume on British Marine Algæ, in which every species that is likely to be met with by ordinary collectors is described, and every British seaweed that is capable of illustra-tion in a work intended for popular information, is figured from plants in my own possession, and, in addition, diagrams and figures from drawings of magnified portions, illustrative of structure and fructification, appear throughout the pages of my work.

(209)

INDEX.

P

London: Printed by W. H. Allen & Co., 13, Waterloo Place, S.W.

* 9 7 8 3 3 3 7 0 2 6 1 8 9 *